# The Biology of Human Survival

# The BIOLOGY of HUMAN SURVIVAL

## Life and Death in Extreme Environments

CLAUDE A. PIANTADOSI, M.D.
Center for Hyperbaric Medicine and Environmental Physiology
Duke University Medical Center
Durham, North Carolina

2003

## OXFORD
UNIVERSITY PRESS

Oxford   New York
Auckland   Bangkok   Buenos Aires   Cape Town   Chennai
Dar es Salaam   Delhi   Hong Kong   Istanbul   Karachi   Kolkata
Kuala Lumpur   Madrid   Melbourne   Mexico City   Mumbai
Nairobi   São Paulo   Shanghai   Taipei   Tokyo   Toronto

Copyright © 2003 by Oxford University Press, Inc.

Published by Oxford University Press, Inc.
198 Madison Avenue, New York, New York, 10016

www.oup-usa.org

Oxford is a registered trademark of Oxford University Press

All rights reserved. No part of this publication may be reproduced,
stored in a retrieval system, or transmitted, in any form or by any means,
electronic, mechanical, photocopying, recording, or otherwise,
without the prior permission of Oxford University Press.

Library of Congress Cataloging-in-Publication Data
Piantadosi, Claude A.
The biology of human survival : life and death in extreme environments /
Claude A. Piantadosi.
p. cm.   Includes bibliographical references and index:
ISBN 0-19-516501-2
1. Extreme environments.   2. Adaptation (Biology)
3. Human physiology.   I. Title.
QP82.P536 2003   612—dc21   2003040497

9 8 7 6 5 4 3 2 1
Printed in the United States of America
on acid-free paper

Adapt or perish, now as ever, is nature's inexorable imperative.
—*H.G. Wells*

# Preface

To persevere across far-ranging environments is profoundly human, but life at the extremes is constrained in extraordinary ways. The diversity of environments in which people are found, either as permanent inhabitants or as temporary visitors, ranges from the high Andes to the scorched Sahara to the frigid Arctic, yet these places are a small fraction of those that harbor life in the thin biosphere around the planet's surface. Most of Earth is too inhospitable for even optimally adapted individuals, and out of necessity, curiosity, or self-indulgence, we have invented technologies to venture into previously impenetrable domains, from the depths of the oceans to the depths of space.

Humans on the frontiers of exploration are tested to the limits of their lives. *The Biology of Human Survival* pinpoints critical factors that dictate life or death at the utmost reaches, including those places accessible to humans only with life-support technology. The book presents environmental physiology using modern, integrated concepts of stress, tolerance, and adaptation. Barriers to life in extreme environments, such as dehydration, starvation, and radiation, are described in separate chapters. Other chapters explain the problems unique to specific environments by examining the determinants of an individual's survival at extremes of cold, heat, altitude, or immersion. Key issues in these specialized settings are illustrated with examples of extreme hardship from great exploits that have attracted people's attention throughout history.

For each environment the book asks these central questions: How does the human body respond to the change in environment and what happens when adaptive mechanisms fail? When does biology reach its limits and when must technology take over? How do scientists evaluate the biological responses to extreme states and solve life-support problems under such conditions? These intriguing questions and their implications offer a fresh look at the human condition.

The book reveals the intricacy with which the human body responds and adapts to environmental change and reminds us that physics and biology collide head-on at many levels, which leads to multiple stresses and numerous opportunities to counter them. As implied by the common etymology of the words *physics* and *physiology*, it is physics that limits life. The physics needed to understand these limitations is explained in language that will be meaningful to students of biology at all levels.

Despite the great heterogeneity of environmental stimuli, all stresses evoke certain common responses. These have been organized in the book to unify general survival principles with mechanisms of adaptation to specific environments. The overarching principles are the body's recognition of stress and the brain's control of physiological systems in order to optimize cardiovascular, respiratory, renal, and hormonal performance. These adjustments conserve and manage vital body resources, such as water, salt, and heat and provide time for the individual to escape or for the body's molecular machinery to adapt.

Probing common reactions to different stresses also provides an opportunity to point out unique stress responses and ingenious solutions to living in marginal environments found throughout the animal kingdom. This allows one to better appreciate why specific functions must be supported in specific environments by man-made devices. Accordingly, the biology in this book is appropriate for engineers and physical scientists as well as any intelligent explorer of the natural world.

The struggle between organism and environment is nature's paradigm, but this is often underplayed in human endeavors. We consider ourselves separate from other animals because we adapt to new environments, in part, by rational action. Thus, the book's underlying theme is the role of behavior in adaptation, emphasizing circumstances in which human technology will forever change the environment, such as after a nuclear war or during colonization of space. Once the subtle interplay of environment with the body's responses and the individual's behavior is grasped, a new window opens onto human survival.

The distinction between biology and behavior is somewhat artificial but conceptually useful. Behavioral adaptation as embodied in modern technology has eliminated most of the day-to-day pressures that molded our ancestors. Virtually instantaneous access to resources such as food, water, shelter, power, medicine, and transportation shape today's individual as much or more than does biological adaptation. The long-term implications of this shift in our way of life are not well understood, and the book describes special environments, such as long voyages in space, in which these unknowns may become especially important.

That environment shapes humanity is never at issue in the book. Rather, we, more than any other species, stand to influence our destiny through our ability to alter the natural world. The natural consequences of global environmental change are familiar to all biologists, and ecological change that creates hardships over which many individuals cannot prevail and multiply will extinguish whole species. A famous example is the effect of the insecticide DDT on the loss of durability of the eggshell of predatory birds, such as the bald eagle (Carson, 1963). Overuse of DDT after World War II threatened their extinction by interfering with the birthrate of hatchlings, a problem that went uncorrected until long after DDT was banned in the United States.

Our propensity to restructure our environment and, soon, our own biology has fantastic implications for human survival that are touched on in the book. This topic has caused theorists to argue over the process of human evolution; some have even proclaimed its end. In any event, modifying the environment at the expense of biological adaptability alters humanity's evolutionary direction. A forewarning of what may await us lies in the fossil record of extinctions brought about by radical fluctuations in climate, but whether change in our environment or our biology is the more significant factor remains unknown.

These matters of "population biology" raise the issue of whether information about our own biology can help us avoid extinction. Human intelligence brings optimism to this prospect, but great cleverness is a double-edged sword that carries the specter of self-annihilation. It is also true that knowledge of human biology is progressing faster than is natural biology itself, but no matter how pleasing the vision of mind over nature, it underestimates natural selection and the effect of the unpredictable on human evolution.

The debate over human evolution is beyond the scope of the book, which deals with the individual, for whom the outcome of environmental stress can be reduced to tolerance and adaptation or death. These outcomes, however, have important ramifications for the long-term survival of humans both on this planet and elsewhere in the solar system. Thus, understanding how individuals adapt to the environment is a step on the road to discovering how the physical world shapes human biology.

*Durham, North Carolina*                                                         C.A.P.

# Contents

1. The Human Environment, 1
   *The nature of human physical boundaries, 1*
   *The importance of preparation for extreme exposures, 4*
   *Some basic concepts of survival analysis, 5*
   *Characteristics of life-support systems, 8*

2. Survival and Adaptation, 10
   *The science of human physiology, 10*
   *Principles of physiological regulation and adaptation, 13*
   *Defining physiological adaptation to the environment, 16*
   *Acclimatization and acclimation, 18*

3. Cross-Acclimation, 21
   *The complexity of adaptation to environment, 21*
   *Positive and negative cross-acclimation, 22*
   *Biochemical mediators of physiological adaptation, 24*
   *Stress proteins and the stress response, 25*

4. Food for Thought, 29
   *A brief overview of human starvation, 29*
   *Starvation: an affliction of the very young and the very old, 30*
   *Assessing the severity of starvation, 31*

*Why children die of starvation, 33*
*Other critical factors in human starvation, 34*
*Starvation and obesity: strange bedfellows, 36*
*The molecular basis of obesity and hunger, 38*

5. Water and Salt, 41
   *The composition of body water, 42*
   *Why do human food and water requirements differ? 43*
   *The body's minimum daily water requirements, 45*
   *The mechanism of dehydration and the body's responses, 47*
   *Dehydration and heat tolerance, 49*
   *Survival time without drinking water, 51*

6. Water That Makes Men Mad, 54
   *The composition of seawater, 55*
   *Ingestion of seawater, 55*
   *Survival at sea, 56*
   *Lessons from the* USS Indianapolis, *57*
   *A practical approach to salt and water loss at sea, 60*

7. Tolerance to Heat, 63
   *Mammalian homeothermy, 63*
   *Humans as tropical primates, 64*
   *Body heat balance, 65*
   *Heat acclimatization, 70*
   *Heat acclimatization and physical fitness, 71*
   *The limitations of human tolerance to heat, 72*
   *Heat illnesses, 73*
   *Death by heatstroke, 76*

8. Endless Oceans of Sand, 78
   *The camel and the Berber, 79*
   *Desert lessons from Pablo and the Haj, 83*
   *Thermal stress and behavior, 84*
   *Importance and regulation of heat-escape activities, 86*

9. Hypothermia, 89
   *The effects of extreme cold on the extremities, 89*
   *Settings for systemic hypothermia, 90*
   *The physiology of hypothermia, 92*
   *Unexpected effects of cold and hypothermia, 95*
   *The subtle effect of winter on human mortality, 96*

10. Life and Death on the Crystal Desert, 99
    *Life in Antarctica, 99*

*The race for the South Pole, 100*
*Failure to adapt to Antarctic conditions, 104*
*Engineering out the need to tolerate cold, 106*
*Human acclimation to cold, 107*
*Estivation, 111*
*Hibernation, 112*
*Hibernation, energy conservation, and suspended animation, 117*

11. Survival in Cold Water, 119
    *The sinking of the* Titanic, *119*
    *Water temperature and human survival, 121*
    *Prediction of survival time in cold water, 121*
    *Survival behavior in cold water, 123*
    *Hypothermia in deep sea diving, 125*
    *Respiratory heat losses and slow cooling, 127*

12. Air as Good as We Deserve, 129
    *Life in an oxidizing atmosphere, 129*
    *Biological oxidations and oxygen toxicity, 132*
    *Antioxidant defenses and the oxidant–antioxidant balance, 135*
    *The free radical theory of aging, 136*

13. Bends and Rapture of the Deep, 140
    *Decompression sickness, 141*
    *Rapture of the deep, 145*
    *Pressure reversal of anesthesia and the high-pressure nervous syndrome, 148*
    *Implications of high pressure for human life on other planets, 150*

14. Sunken Submarines, 152
    *The sinking of the* Kursk, *152*
    *The debate over submarine escape, 155*
    *The physics of submarine disasters, 156*
    *Analysis of survival factors on sunken submarines, 158*

15. Climbing Higher, 164
    *The physical environment of high altitude, 164*
    *Physiological responses to high altitude, 166*
    *High-altitude illnesses, 173*
    *The zone of death, 176*
    *Limits of human ascent to high altitude, 179*

16. Into the Wild Blue Yonder, 181
    *The International Standard Atmosphere, 181*
    *Human visitation to the stratosphere, 183*
    *Depressurization accidents, 185*

*The Armstrong line, 188*
*The pressure suit, 189*

17. G Whiz, 193
    *The continuity principle, 193*
    *Gravity and acceleration, 194*
    *High-G environments, 195*
    *Limits of high-G tolerance, 197*
    *Adaptation to sustained G forces, 201*

18. The Gravity of Microgravity, 203
    *Space sickness, 204*
    *Intolerance of upright posture, 204*
    *Loss of bone mass in space, 206*
    *Loss of muscle mass in space, 209*

19. Weapons of Mass Destruction, 212
    *Biological and chemical warfare agents, 213*
    *Thermonuclear weapons, 217*
    *Types of radiation, 219*
    *Biological effects of radiation, 220*
    *Radiation and the human body, 223*

20. Human Prospects for Colonizing Space, 227
    *Advanced life-support systems, 228*
    *Mission to Mars, 230*
    *Habitability factors in long-duration spaceflight, 232*
    *Deleterious effects of long-term exposure to microgravity, 234*
    *Effects of life in space on human immunity, 235*
    *Long-term effects of radiation on human life in space, 238*
    *Establishment of permanent human populations in space, 242*

    Bibliography and Supplemental Reading, 247

    Index, 255

# The Biology of Human Survival

# 1

# The Human Environment

As with any other species, human survival boils down to individual survival. This is true whether people die of disease, natural disaster, or manmade holocaust. Fundamentally, survival can be defined in terms of the interactions between an individual and its natural surroundings. The surroundings determine the extent to which a person is exposed to critical changes in environment, such as temperature, water, food, or oxygen. The physical world imposes strict limits on human biology, and learning where these limits are and how to deal with them is what biologists call *limit physiology*. The principles of limit physiology can be applied to understanding human life in all extreme environments. These principles will be developed in this chapter and applied throughout the book to gain a deeper appreciation of how humans survive in extreme conditions.

## The Nature of Human Physical Boundaries

One of the most important characteristics of every living organism is its ability to maintain an active equilibrium, however brief or delicate, with its natural environment. All living beings, as integral parts of nature, can be characterized by the dynamic exchange they maintain with their physical surroundings. Being alive requires being attuned to natural change, and many organisms are exquisitely sensitive to even tiny perturbations in environmental conditions. They occupy restricted *niches*. If changes in conditions in the niche exceed certain limits, biological equilibrium is disrupted, and the life of the organism, or even the entire species, is threatened. Thus, all habitable environments, or *habitats*, have specific physical boundaries within which life is possible and outside which life is impossible. As an organism approaches the limits of its habitat, life is sustainable only with greater and greater effort unless the effort is sufficient for adaptation to occur. Indeed, the closer the organism approaches a tolerance limit, and the greater the stress, the more vigorous will be the attempt to compensate, and if it falls short, the shorter will be the survival time. This principle is illustrated in Figure 1.1. The curve has the shape of a rectangular hyperbola, which is characteristic of many survival functions depicted throughout this book.

Human beings are among the most adaptable creatures on the planet, yet the limits of human survival are astonishingly narrow when viewed in the context of the extremes on the planet. Approximately two-thirds of the Earth's surface is covered by deep saltwater oceans, which air-breathing terrestrial mammals such as ourselves may visit briefly but are not free to inhabit. Even highly specialized diving mammals, the great cetaceans, so spectacularly adapted for life in the sea, are confined to the surface layers of the ocean. The crushing pressure of the seawater, the cold, and the darkness make the great depths of the ocean inhospitable to most marine species. Not that life cannot exist or even thrive under such extremes, for even at the bottom of the sea super-hot water jets heated by vents in the Earth's mantle sup-

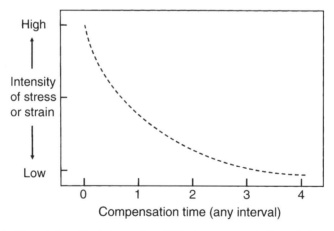

**Figure 1.1.** The relationship between the ability to compensate for and the severity of physiological stress or strain. X-axis indicates time to failure of function, or, in the case of survival, to death. The time scale may be in any unit, from seconds to days, depending on the nature and intensity of the stress or strain. Adaptation shifts the position of the curve to the right.

port highly sophisticated and unique forms of life, but the thought of people existing permanently in such places is unimaginable. Many species that thrive in the depths die when brought too quickly to the ocean's surface.

Of the land that covers the remaining third of the Earth's surface, one-fourth is permanently frozen and one-fourth is arid desert; both are extremes that may be inhabited by humans only with arduous efforts. Add to this the high mountain ranges and the lakes and rivers that people depend on but do not routinely inhabit, and the climate and topography temperate enough for permanent habitation by humans relegates us to one-sixth of the surface of the planet. Despite the rich diversity and capacity of people to respond, adapt, or acclimatize to extreme conditions, the limits of human tolerance are remarkably narrow. Indeed, civilization's stamp has been its ability to extend an individual's tolerance to stressful environments by making behavioral adaptations in the form of invention (Figure 1.2).

The environment of human beings is constrained geographically because latitude and topography cause variations in temperature, barometric pressure, availability of food and water, and combinations of each that are critical for survival. Thus, it is no surprise that much of the world's population is perched on the brink of disaster. The loss of human life from a sudden natural disaster such as a blizzard or a flood is always appalling, but it is remarkabe how well some individuals

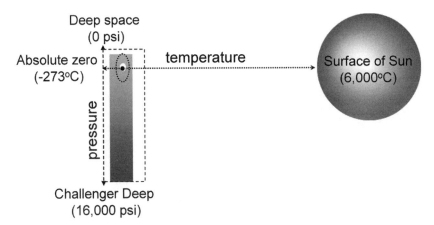

**Figure 1.2.** The physical environment of human beings. Natural environments are indicated on absolute temperature (X) and pressure (Y) scales. The shaded rectangle shows natural physical environments of life on Earth. The small black spot in the center indicates the range of the human natural environment, and the white circle is the region of extension of tolerance by physiological adaptation. The dotted circle indicates the range of tolerance by inventions designed to support or protect one or more critical body functions during an exposure, and the dashed line is the range of hard shell engineering, which prevents exposure to an extreme external environment. The latter two responses are forms of advanced behavioral adaptation.

endure the most grueling conditions, surviving prolonged immersion, high altitude, and heat or cold despite desperate thirst or impending starvation. The incredible tales of survival at sea, in the mountains, in the desert, and on the ice capture our imaginations like few others.

## The Importance of Preparation for Extreme Exposures

Every life-and-death struggle is influenced by intangibles, sometimes lumped under broad terms such as *survival instinct* and *will to live*. Whether an individual survives an unexpected and prolonged encounter with a potentially lethal environment, however, depends more on the equilibrium between biology and physics than on intangibles. Although much has been made of a strong will to live, this is a basic trait of the human psyche common to healthy people. Strength of spirit, motivation, and psychological factors are very important for survival but are less decisive under truly catastrophic conditions than our poets and writers would like us to believe. To state it plainly, rarely does one person survive under extreme conditions when another dies simply because the survivor has a greater will to live.

Thoughtful preparation in anticipation of extreme exposure is more important than all the fighting spirit in the world, for a naked man cannot live out a night at the South Pole. Preparation, however, requires knowledge, time, and resources. It involves allowing time to adapt, for example by gradual ascent to an altitude or by arranging resources to limit the effects of the exposure, such as by providing multiple layers of warm, dry clothing on polar expeditions. Preparation when an automobile breaks down in the desert means simply avoiding death from dehydration by having had enough foresight to carry along some water. This example implies the double failure that has killed many a bold explorer. One failure occurs before the adventure begins by counting on a single vehicle and not carrying enough water to walk out or to survive until another vehicle can come to the rescue. The second failure, engine trouble, usually nothing more than an inconvenience, proves fatal.

The double failure problem is well known to engineers who design life-support equipment such as diving gear and spacesuits. They devise diagrams to analyze potential failures or faults in systems that will affect the probability of survival in specific failure modes. These fault analysis diagrams, or trees, can become quite complex for even relatively straightforward systems. However, most of the essential information can be gleaned from simple diagrams, if properly constructed, and it is surprising how few explorers actually use this approach in planning an expedition. Fortunately, the prudent explorer appreciates the bottom line: the way to ensure safety and dependability is to build in redundancy. Deciding how much redundancy is enough is the tough part.

## Some Basic Concepts of Survival Analysis

The double failure problem and the value of redundancy can be illustrated with diagrams using a technique called nodal condensation probability. In planning an expedition, if one anticipates event A, a 1-in-1000 failure with a 99% chance of survival, the probability of death is only 1 in 100,000. These odds are acceptable to most people. However, if a second independent failure possibility, B, is added to the expedition with the same probability and the same survival rate, the two probabilities must be summed, giving an expected risk of death of 1 in 50,000 for the expedition. This arrangement of events, known as a linear system, is depicted in Figure 1.3.

One must also consider the effect of a rare double failure because the odds of surviving the second failure may approach zero if it occurs after the first failure.

Start ──(**a**)──(**b**)──▶ Finish

$P_a = 1{:}1{,}000$   $P_b = 1{:}1{,}000$   $P_{ab} = 1{:}1{,}000{,}000$
$P_d = 0.01$   $P_d = 0.01$   $P_d = 1.0$
$P_{d(a)} = 1{:}100{,}000$   $P_{d(b)} = 1{:}100{,}000$   $P_{d(ab)} = 1{:}1{,}000{,}000$

$$P_{death(tot)} = P_{d(a)} + P_{d(b)} + P_{d(ab)}$$
$$= 2.1{:}100{,}000$$

If **a** fails before start:

$P_a = 1$   $P_b = 1{:}1{,}000$
$P_d = 0$   $P_d = 1.0$
$P_{d(a)} = 0$   $P_{d(ab)} = 1{:}1{,}000$

$$P_{death(tot)} = P_{d(ab)} = 1{:}1{,}000$$

**Figure 1.3.** Survival probability by linear failure analysis. The top part of the diagram shows two independent events, **a** and **b**, in a linear arrangement. The probability (P) of each event is 1 in 1000, and the probability of death ($P_d$) if one event occurs is 1:100. The probability of death for each event is therefore the product of $P \times P_d$, or 1 in 100,000. The probability of both events occurring is the product of their probabilities, or 1 in 1 million, but if both occur the probability of death is (set at) 1.0 (certainty). Therefore, the overall risk of death is the sum of the three products, or 2.1 in 100,000. The bottom part of the diagram shows two independent events, **a** and **b**, arranged in a linear system in which **a** has already occurred but has no consequences because it occurs before **b** in a different environment or location, for example, before an exposure. However, if the exposure is undertaken and **b** then occurs, the probability of death increases from 1 in 100 to 1.0 (certainty) because **a** is already in place. Therefore, failure to account for **a** fixes the probability of death at 1 in 1000, which is nearly fiftyfold higher than in the top part of the diagram. Examples are provided in the text.

A good example is ejecting from a burning jet aircraft with a defective explosive canopy bolt. If an independent probability of 1:1000 is assigned to each event, the probability of experiencing the double failure is 1:1000 squared, or only one in a million. The chances of living through it however, are virtually zero. Overall, the expected probability of death in this linear system is 2.1:100,000.

Next, consider the problem of starting an expedition with a failure already in place. This is illustrated in the bottom half of Figure 1.3. In the desert example above, the motorist left town without a supply of drinking water. In this case, the event, $a$, is assigned a probability of 1 because it happened, but it did not happen in the desert and the motorist can find water anytime before departure. The probability of dying of dehydration is nil. The second failure, $b$ in Figure 1.3, has a probability of 1:1000, but now the probability of death is 1. This means the overall risk of dying on the expedition has gone from 2.1:100,000 to 1:1000, nearly a fiftyfold increase. These calculations give one an appreciation for why most of the deaths in mountaineering, deep sea diving, parachuting, and so on are due to double failures that involve at least one human error. The initial failure often encompasses a critical failure of preparation.

Probability calculations illustrate the value of assessing risk and preparing in advance for a trip of significant intrinsic danger. The importance of redundancy to reduce danger, although intuitive to most, can be made clear with examples. For instance, underwater divers who explore caves carry both extra lights and an independent breathing gas system. This greatly lessens the chances of double failure, such as no light and no air, and improves survival after the potentially critical failure of getting lost, for example by dropping or becoming disconnected from the lifeline.

Another example of redundancy is the use of personal flotation devices and safety harnesses on ocean-going sailing yachts. If a member of the crew falls overboard wearing a life jacket, it supports him or her until the boat comes about to make the pick up. Hence, the probability of death is quite low. However, this safety measure is not sufficient under all conditions. If the person is alone on deck and falls overboard, the probability of death is very high despite the life jacket because the boat will sail away faster than he or she can swim. Harnessing one's vest to the vessel beforehand decreases the probability of death from falling off the boat under special circumstances, such as standing watch alone at night. This principle of redundancy is illustrated in Figure 1.4.

In many extreme situations strength and toughness may have appeared to swing the odds in favor of survival, but analysis of the events usually indicates this was because some deadly factor was held at bay by rational actions. In dramatic, highly publicized examples of survival against extreme odds, an injured party is snatched from the jaws of death in the nick of time. The would-be victim and the courage of the rescuers are applauded by all, and justifiably so, for this is the stuff of leg-

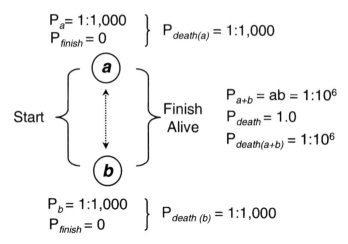

**Figure 1.4.** Survival probability in a parallel system. In this diagram events **a** and **b** are arranged in parallel. In other words, if **a** occurs, option **b** can be exercised to prevent death because of **a**, and vice versa. Thus, double failures, **a** and **b**, must occur to cause death. The risk of death during the exposure is greatly reduced, from 1 in 1000 to 1 in 1 million, by the redundancy. Examples are provided in the text.

ends, yet survivors are often the best-prepared and most knowledgeable individuals in harm's way.

Whether someone lives through extreme exposure can be boiled down to a few physical and biological factors. In its minimal form survival analysis requires an accounting of four factors, which can be defined as critical variables. The first two variables are beyond human control, while the latter two are amenable to intervention. These critical variables are as follows: (1) the physics of the environment, (2) the limits of human physiology, (3) the length of the exposure, and (4) behavioral adaptation, including what the victim understands about survival requirements and the plans made to prepare for a failure.

This approach simplifies the analysis, but not greatly because the four critical variables are complex. In other words, they are true variables, neither constants nor necessarily simple changes, and this makes survival prediction an inexact science. For instance, hostile environments do not produce "pure" physiological stresses; many places are both hot and dry, such as the Sahara, or cold and high, such as the Antarctic. This results in multiple stresses on the body that interact with one another. To further complicate the situation, human biology encompasses differences in body shape, mass, and fitness that greatly influence survival time under different conditions. This aspect of the problem, known as physical diversity, is most obvious for survival in cold water. Physical diversity implies that certain body characteristics, such as, fatness, carry different degrees of importance under different conditions, such as providing temporary advantage while immersed in cold water.

As a general rule, the order in which the factors are listed above is their order of importance, even if parameters within each factor change. On the other hand, when a potentially lethal exposure is in the offing, the fourth factor, behavioral adaptation made by interventions by victim and rescuer, is the only means of producing a survivor. The principle also holds when one prepares for extreme exposure known to exceed one of the body's physiological limits. The only effective survival strategy is to use intelligence and a priori knowledge, that is, behavioral adaptation.

Some readers may take issue with this ordering of the variables or this approach in general. Even so, it must be admitted that many places on Earth are too hot or too cold, or the pressure too high or too low, to permit unassisted human survival long enough for adaptation to occur. These places, wherever they are, define the limits. Therefore, the chapters in this book, although intended to tease out essential commonalities of human survival, are organized according to particular environments that place unusual demands on the human body.

In order to understand the limits of life in these environments, some working concepts must be provided for the biology of the human organism. This will require definition of the essential tenets of modern environmental physiology including the concepts of homeostasis and adaptation which will follow later. Homeostasis and adaptation also require support from the environment, such as food and water. Conditions devoid of food or water constitute special environments of hunger and thirst that lead to foraging and ultimately, if unsuccessful, to starvation or dehydration. These factors in turn impact survival in special physical environments where people encounter extremes of temperature, barometric pressure, radiation, or gravity.

## Characteristics of Life-Support Systems

Survival analysis embraces the principles of life-support equipment. Life-support problems are encountered whenever physiologists and engineers collaborate to fashion systems to support human beings in extreme environments. The issues are similar for systems as simple as a diver's wet suit or as complex as the International Space Station. Not surprisingly, designs hinge on just how closely human beings should be allowed to approach a biological limit. In general, the more hostile the external environment and the closer the internal environment is kept to natural conditions, the greater the engineering requirements and the higher the cost. In considering life-support equipment, three distinct but related environments are always involved: the internal environment of the body, the environment adjacent to the body, and the external environment, that is, the environment outside the suit or system. By first principles, the objective is to maintain the internal stability and functions of the body, which means that the critical environment is immediately adjacent to the body. Thus, in deep space, where air pressure is virtually zero, a spacesuit is worthless if it cannot maintain an internal pressure compatible with physiological activity. Critical body functions will be compromised unless the pressure in the suit is equivalent to 30,000 feet of altitude or less.

To make the concepts of life support accessible, the discussions of different environments rely only sparingly on mathematics, which is limited to a few equations that offer exceptional insight into the relationships between the human body and its environment. The algebra helps add clarity to explanations of the principles of life-support technology. Many brilliant and innovative breakthroughs in life-support technology were made in the twentieth century, but mathematical models and technological subtlety are not usually critical to understanding the principles that most directly limit survival in each environment and what happens to people when equipment fails.

The principles of life support encompass many human problems, including disease, fitness, isolation and group dynamics, circadian rhythms, and sleep deprivation. These fascinating topics are important not only in extreme environments but may be crucial to life in artificial systems. They are not neglected in this book, but their effects are discussed in association with the appropriate physical environment. This approach places them in context and highlights some counter-intuitive notions, for example, that physical fitness and the ability to perform work offer a survival advantage in some environments but not in others. The ultimate extreme environment, that of space, is used to point out gaps in our knowledge of long-term human endurance and adaptation outside the confines of Earth. Throughout the book common determinants of survival in artificial environments are highlighted as much as possible, together with their implications for the future of humankind.

# 2

# Survival and Adaptation

The relationship between any living organism and the environment in which it lives is as intricate and delicate as is the relationship between a species and its ecosystem. Highly complex biological systems such as the human body coordinate many processes to maintain a stable internal environment despite fluctuations in the environment around them. The integrated regulation of these processes is designed to allow the organism to function properly amid the varying influences of the external environment. Integration is the key principle in understanding how the human body functions.

## The Science of Human Physiology

The study of how a living organism functions normally despite disturbances in the environment is the discipline of physiology. Because all living systems attempt to maintain constant internal conditions in the face of changes in the environment, it can be argued that all physiology is essentially environmental. This view was first espoused by the scientist–theoretician Claude Bernard at the Sorbonne in the second half of the nineteenth century. Indeed, Bernard, who made many seminal discoveries in experimental physiology, is undoubtedly most famous for his concept of the *milieu-intérieur* of the body (Grande and Visscher, 1967). His simple but invaluable generalization states that the body fluids of higher animals form an internal environment that provides the conditions needed to sustain life at the level of its fundamental units, such as the cells and tissues. Bernard also knew that the integration of functions was accomplished by the workings of the blood, circulation, and nervous system.

Having drawn a distinction between the internal and external environments of higher organisms, Bernard went on to point out that these two environments are independent of each other. The self-organizing structure of "continuous life," as he called it, enables organisms to maintain the internal conditions necessary for continuous function via physiological processes such as respiration, circulation, water and salt balance, transfer of heat, and storage of nutrients. The concept of the *milieu-intérieur* eventually came to epitomize modern physiology, which until Bernard's time had been considered natural philosophy, not medical science.

Many of Bernard's students made important contributions to environmental physiology, including Paul Bert, his direct successor to the chair in general physiology in the faculty of sciences at the Collège de France. Bert remains known today for his contributions to respiratory physiology and the effects of altered barometric pressures and oxygen on the body. His principal work, *La Pression Barométrique*, published in 1878, is a historical landmark in environmental physiology. At the time, severe adverse and sometimes lethal effects of high altitude had been encountered in balloonists, but their cause was a scientific mystery. Bert correctly connected the deleterious effects of high altitude to a lack of oxygen ($O_2$). He also correctly predicted the barometric pressure on the summit of Mount Everest and exposed himself to that pressure in a vacuum chamber while breathing oxygen. Bert's remarkable book was reprinted during World War II by the U.S. Army Air Corps to acknowledge his pioneering role in aviation medicine (Bert, 1943).

In addition to Bert, Bernard's thinking influenced many of the great physiologists of the early half of the twentieth century, including Joseph Barcroft and J. S. Haldane in England. Barcroft, who lived from 1872 to 1947, spent most of his scientific career at Cambridge working with one molecule, hemoglobin, in order to understand its ability to combine reversibly with $O_2$. Hemoglobin's ability to bind and release $O_2$, together with lungs or gills to extract $O_2$ from air or water and a sophisticated cardiovascular system with its miles of tiny capillaries to deliver oxygenated hemoglobin to the cells, is the primary reason that large, mobile animals are able to exist (see Fig. 2.1). Indeed, the principle of nutrient circulation of blood to tissues had been recognized since 1628, when English physician William Harvey published his famous treatise, "On the Motion of the Heart and Blood in Animals."

Joseph Barcroft is most widely remembered for his studies on lack of $O_2$, or hypoxia (Barcroft, 1914). He subsequently classified the problem of hypoxia into three categories. Barcroft recognized that the supply of $O_2$ needed for life could be compromised by more than just failure of breathing, or external respiration. Hypoxia could arise from insufficient $O_2$ in the blood (hypoxic hypoxia), lack of

12  THE BIOLOGY OF HUMAN SURVIVAL

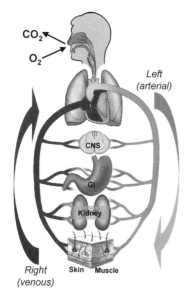

| Major Physiological Functions of the Body ||| 
|---|---|---|
| Organ System | Principal Function | BMR (% Total) |
| Lungs | Gas Exchange | 2% |
| Heart | Blood Circulation | 15% |
| Central Nervous System (CNS) | Integration of Systems | 20% |
| Gastrointestinal (GI) | Nutrition | 15% |
| Kidneys | Excretion Fluid Regulation | 10% |
| Skeletal Muscle | Work Thermogenesis | 35% |
| Skin | Barrier Thermoregulation | 1% |

**Figure 2.1.** The major physiological functions of the body. Five major organ systems of the body are shown in the framework of the pulmonary and circulatory systems. The table indicates the main functions and the percentage of basal metabolism (BMR) necessary to support function. Overall metabolism increases and the distribution of metabolic requirements change when work is performed.

hemoglobin in the blood (anemic hypoxia), or inadequate flow of blood to tissues (stagnant, or ischemic, hypoxia). About ten years later a fourth type of hypoxia was added to the list: poisoning of internal respiration, cytotoxic (or histotoxic) hypoxia, such as that caused by cyanide. Barcroft's classification of hypoxia is still taught today.

Barcroft's concepts are even more remarkable in the context of the scientific ambience of his day. Since Lazaro Spallanzani (1729–1799), the idea that living tissues consumed $O_2$ had been considered, but this was disputed even by Bernard, who believed $O_2$ was consumed in the blood. To those who recognized cellular respiration, such as the Nobel laureate Otto Warburg, the source of $O_2$ consumption was unknown. Warburg thought there must be a system to reduce $O_2$ in tissues, which he called *Atsmungferment*, but not until 1925, when David Keilin (1887–1963) published his work, were chemical pigments in the cell, mitochondrial cytochromes, revealed to be the source of respiration (Keilin, 1966).

The other giant of British physiology of the early twentieth century, J. S. Haldane, is perhaps most widely known for publishing in 1908 the first tables on safe decompression for use by compressed air divers. Haldane was a brilliant and innovative scientist who made many other important contributions to physiology. For instance, he recognized that many of the toxic effects of carbon monoxide (CO) on the body are due to lack of $O_2$ and uptake of CO by tissues. It was Haldane who proposed using the canary, which is rapidly poisoned by CO, to warn miners of the presence of toxic coal gas in tunnel atmospheres.

In the United States the great Harvard physiologist Walter B. Cannon (1871–1945), famous for the concept of homeostasis, paid tribute to Bernard's *milieu-intérieur* in the development of his own ideas. In the preface to the French edition of his book *The Wisdom of the Body*, Cannon acknowledged that its main tenet, that the stability of the inner medium of the organism is actively regulated in higher vertebrates, was "directly inspired by the precise view and deep understanding of the eminent French physiologist Claude Bernard (Cannon, 1932)." Cannon's contribution to environmental physiology is often overshadowed by his monumental contributions to general physiology. Nonetheless, he was the first to clearly articulate the idea that failure of homeostasis leads to death when the organism is exposed to extreme environments.

## Principles of Physiological Regulation and Adaptation

The ability to maintain homeostasis despite changes in the environment that would otherwise disturb it is the hallmark of adaptation. The word *adaptation* has a certain ambiguity, but for the biologist it defines the extent to which an organism can occupy an environment, use available resources, and multiply. A biologically successful species broadens its range by adaptation in accordance with the principles of evolution and population genetics. To paraphrase René Dubos, this simple definition does not do justice to modern humans, whose behavior (influenced by socioeconomic and cultural factors) has become a principal driving force in adaptation (Dubos, 1965).

In the late 1940s the scientific pioneer E. F. Adolph described physiological regulation in terms of changes in body function caused by changes in the environment. Adolph was one of the first to recognize clearly that physiological regulation is directly responsible for the ability to adapt to a new environment or conditions (Adolf, 1956). Physiological regulation provides the signals and the time necessary for the normal processes of adaptation to occur.

Although the concept of homeostasis is clearly essential to understand normal physiology, it is also important in understanding disease, whereby exaggerated or attenuated homeostatic responses occur as a result of the disease process. The roles the body's compensatory mechanisms play in disease are interwoven into the astonishingly intricate processes of pathophysiology. Disease, by definition, disturbs the homeostasis in some way, and the body "adapts" with an apparent homeostatic response. Thus, a pathophysiological response may resolve one disorder while creating another. As will also become apparent, some "adaptations" are actually maladaptive.

Modern biologists operate within a narrower frame of reference than that of the homeostatic mechanisms of a whole organism. The body is reduced to a group of systems, such as the cardiovascular system, each containing a set of regulatory and functional components. The regulatory components are often described by specialized and sometimes arcane terms, such as *set point, negative feedback loop*, and *gain*. These terms are conceptual tools for describing different aspects of homeostasis. The basic components of each system are essentially the same: a parameter to be regulated, a detector to measure the parameter, a desired value (the set point), and a comparator to measure the difference between the set point and some reference value. The comparator computes the difference between the set point and the reference value and generates an error signal that initiates a compensatory response. The intensity of the error signal determines the intensity of the compensatory response. This is the gain in the system. The gain determines how rapidly the parameter returns toward the set point, thus completing a feedback loop. As the actual value of the parameter approaches the set point, the error signal becomes smaller and the intensity of the compensatory response diminishes.

An excellent example of a regulatory system is provided by mammalian temperature regulation, or thermoregulation (Benzinger, 1969). In its simplest form, body temperature may be thought of as the regulated parameter, with a desired value, or set point, of 37°C. When the body is exposed to cold air, it loses heat, skin temperature falls, and an increase is detected in the difference between skin temperature and body temperature. An error signal is generated, and shivering begins. Shivering produces extra heat in an attempt to maintain the body temperature at 37°C. If shivering is not vigorous enough, body temperature will begin to fall, the difference between the set point and actual body temperature will be detected, and shivering will intensify. In this example two reference values are used, skin temperature and body temperature, or the set point itself. Nothing specific need be said about the nature of the detectors or the comparators, but their presence is implied by the response. In some instances specific biological structures have been identified as detectors and comparators, but in many physiological systems these sensors remain incompletely understood.

Such model systems are obviously oversimplified, yet they are adequate to illustrate the main elements of physiological regulation and the key forces of adaptation. However, as is so often the case, the devil is in the details, and the primary response is typically modified by more subtle events that produce nuances and complexities. Such complexities confound most models of biological systems, particularly with respect to thresholds for responses, inertia in the system, interactions among different parameters and signals, and the nature of the behavioral responses involved, yet simple models often provide remarkable insight about what to expect of the body in terms of adaptation after a change in external environment. Model building points out ways to design further scientific experiments that lead to better models.

Models of physiological systems based on set points and negative feedback loops have been conceptually useful in understanding homeostasis, but models fail to explain how homeostasis is maintained over the long run, particularly because even a simple organism is capable of modifying its behavior in anticipation of the effects of continuing stress. I will return to this idea, but it has been pointed out that set points need not exist for a system to show regulated behavior. Rather, a system or set of systems may simply "settle out" at a new equilibrium value in accordance with the change in external environment (or internal environment in the case of compensation for disease). Thus, control is exerted around settling points that vary somewhat with the exact conditions. This idea is very attractive because it allows for the range of normal values that are observed in nature.

A good way to illustrate the settling point principle in human physiology is to consider the effects of a chronic disease on a single organ system. The disease damages the organ, and it begins to fail in its function. Nevertheless, the organism usually does not die until the damage becomes very severe; first, the parameters regulated by the organ system change and begin to fall outside the normal range. This allows an observer, such as a physician, to follow the progression of the disease.

There are many examples of this principle appearing to operate in the body, such as regulation of blood glucose by the pancreas in diabetes. Another particularly illustrative example is the effect of high blood pressure on the kidneys. If high blood pressure is sustained for many months (hypertension), stress on the tiny blood vessels in the kidneys will gradually cause the cells that make up their smaller functional units to begin to die. If blood pressure is not brought under control, damage to the kidneys causes further increases in blood pressure, and eventually the kidneys will fail completely. The person will require dialysis. The physician, aware of this problem while treating the hypertension, will follow the levels of urea nitrogen and creatinine in the blood, which are metabolic waste products normally excreted by the kidneys. As the kidneys fail, urea accumulates in the blood, and a new equilibrium is reached at a higher level of urea. Thus, the system settles out at a different point. As long as the urea is not too high, it is well tolerated, and there are no significant physiological problems. Eventually, however, kidney function will decline to the point that the accumulation of urea and other metabolites will interfere with brain function and result in the coma of uremia.

The problem of rising blood urea could be viewed as a change in the set point for the clearance of urea by the kidneys. However, this view of the system begs the question because if something intervenes to reverse the hypertension, recovery of kidney function may occur, and the urea will come down and may even settle into the normal range. A set point scheme would require constant resetting of the set point. Here this idea becomes rather cumbersome. Thus, the usefulness of set point modeling varies with the system of interest.

There is also a problem with the idea of negative feedback loops that respond to error signals. For one thing, error signals can be very difficult, if not impossible, to identify. In other words, many physiological systems respond precisely to a stimulus before an error signal is detected. One may simply argue that the tool being used to measure the error signal is not sensitive enough, but something is missing in this way of thinking about physiology. Living organisms change the actions they take in response to a change in the environment simply on the basis of the change. For instance, a litter of kittens huddles together in the cold before the body temperature of any one kitten begins to fall and it begins to shiver. In addition, organisms with sophisticated nervous systems can respond to a change in environment based on past experience. In other words, they learn to avoid a noxious place and to anticipate future needs for food, water, or protection from the elements. These topics will be explored in more detail in discussing behavioral adaptation.

## Defining Physiological Adaptation to the Environment

An appropriate vocabulary is needed to think and communicate clearly about how living organisms respond to their environments. Biologists speak freely of adaptation, acclimatization, acclimation, deacclimation, accommodation, and habituation. Because conceptualization of biological principles relies rather heavily on a common terminology, it is important to describe these terms as precisely as possible. The general terms are defined first, followed by more specific terms that will be useful later.

Because every biological event is a response to changing conditions, each change in a body function reflects physiological regulation. Body responses as diverse as changes in blood pressure, metabolic rate, defensive behavior, and reproduction are all regulated events that occur because conditions change. Therefore, any response designed to allow an organism or a species to survive represents a form of adaptation. If adaptation is defined in this way, it is easy to recognize that adaptive processes are themselves physiological responses. Reflecting on this in the context of a population, it is apparent that adaptations take one of two forms. One form occurs in individuals and the other in individuals' offspring. These forms of adaptation are referred to as physiological and genetic adaptation, respectively.

Physiological adaptation can be defined most simply as any functional, structural or molecular change that occurs in an individual as a result of exposure to change in the environment. During physiological adaptation the individual avoids certain conditions that have caused problems in the past. This simple definition avoids categorizing adaptation by intensity, rate of onset, duration, and sequence of responses. For instance, physiological adaptations may develop and be completed within milliseconds, or they may require months. This definition avoids making any statement about reversibility; adaptive responses may be rapidly reversible, slowly reversibly, or irreversible.

A broad definition of physiological adaptation also encompasses genetic responses in the individual, that is, genetic changes in somatic cells that may permanently alter individual fitness (phenotype) but that cannot be passed on to the individual's progeny. Changes in reproductive or germ cells caused by environmental factors *can* be passed on to the offspring, but these changes are genetic *mutations*. Mutations may be adaptive, neutral, or maladaptive but ultimately are responsible for genetic adaptation and evolution.

Genetic adaptation is most simply defined as a structural or functional change built into the molecular genetic code of a species or strain of organisms that favors survival in a particular environment. This summary definition of genetic adaptation includes two very specific features. It requires that adaptation occur from a permanent change in the germ cell line of an individual. Thus, the adaptive trait is heritable; it is a permanent genetic mutation. As a result of this, genetic adaptations require a great deal of time; in fact, they usually require many generations to spread through a population. The time, or generational requirement, for this form of adaptation is closely linked to the second feature, the process of natural selection. Natural selection is the only known mechanism whereby a survival advantage conferred by a specific genetic change can be transferred to an entire population of organisms in a new or changed environment.

A spontaneous, or de novo, mutation, which may or may not have an environmental cause, can propagate rapidly throughout a species if it imparts a survival or reproductive advantage. Neutral mutations tend to propagate more slowly for the obvious reason that they lack advantage. Deleterious mutations also propagate slowly and generally do not extinguish a species because the individuals in whom they occur tend to be deselected for reproduction. Clearly, however, modern medicine has changed this operating principle for human society. Another principle is also important: a mutation that imparts a survival advantage in one environment but a disadvantage in another can extinguish an entire species if the first environment converts to the second.

The principle of this kind of interaction between environment and molecular genetics can be illustrated by the disease scurvy, which is caused by ascorbic acid deficiency. Ascorbic acid, or vitamin C, is not required in the diets of many mammals because their bodies synthesize it. However, humans presumably lost the ability (an enzyme) to synthesize ascorbic acid, and we must obtain vitamin C from our diets. As long as fresh fruits and vegetables are available, scurvy does not exist, but if vitamin C cannot be supplied from the environment, scurvy emerges and can become a fatal disease. This simple principle was proven in the seventeenth century by the British physician Joseph Linde in British seamen in the first controlled experiment in human nutrition.

The specific advantages of physiological adaptation for survival of individual human beings are emphasized throughout this book, but a good deal of what is covered has staggering implications for genetic adaptation in the future. Some of these implications are related to the process of acclimatization, as defined in the next section. Long-term physiological adaptation to new environments involves both critical and noncritical factors, and human interventions, such as life-support design, are likely to identify and support critical factors known to be necessary for individual survival. As a result, the presence of heightened, diminished, or entirely new physical forces on many generations of people may influence the genetic composition of offspring. By natural selection, such forces can confer permanent traits with new survival advantages and perhaps eliminate some of those that are no longer needed. As a result, the biology of future generations of human beings, who may be suited to life on the Moon or in deep space, may be very different indeed from present-day people.

Permanent adaptations are built into the genetic code (genotype) of individuals who emerge from less well-equipped populations and presumably offer an advantage over temporary processes, such as acclimation. Whether permanent adaptations appear de novo or arise gradually from the capacity to acclimatize does not matter in the context of the extraordinary breadth of possibilities given time and the right circumstances.

## Acclimatization and Acclimation

The process of adaptation that occurs over a period of days to months in response to a change in natural environment is known as acclimatization. For example, individuals generally adapt to life in the desert or at high altitude by acclimatizing to all the features of the new environment. Because changes in the natural environment usually involve more than one physical process, for example, both temperature and altitude on a trip into the Himalayas, acclimatization involves adaptation to two or more environmental factors. Acclimatization differs from acclimation because it involves responses to complex environments rather than to a single environmental condition. Thus, acclimatization results from the interaction or summation of the effects of two or more environmental factors on the responses of the body.

Acclimation involves physiological adaptation to a single environmental factor, or stressor, such as a change in environmental temperature or a change in altitude. These stressors are also known as adaptagents. When a stressor is sufficiently intense to invoke a biological response, such as by exceeding some threshold, the stimulus–response is referred to as strain. Conventional theory holds that only stressors that produce strain will result in adaptation. Independent changes in single stressors, or adaptagents, are rarely, if ever, encountered in nature, but they can be simulated in artificial environments. Studies of acclimation are conducted in environmental chambers in which one stressor at a time is manipulated and the others are tightly controlled. Such artificial manipulation of adaptagents is scientifically valuable but may produce physiological responses quite unlike those encountered when the stressor occurs as part of a change in the natural environment.

To some biologists the terms *acclimatization* and *acclimation* imply cause-and-effect relationships in which the effect develops rather slowly, but some adaptations can occur very quickly following a stress and reverse almost as rapidly when the stress is removed. These forms of adaptation are known as accommodation. For example, the pupil of the eye accommodates rapidly to changes in light intensity: it dilates in response to darkness and constricts in response to sunlight. However, not all types of accommodation reverse rapidly when the stressor is removed; some require minutes or hours before returning to normal. Thus, drawing a distinction between accommodation and acclimation on the basis of time has not proven particularly useful. It is more useful to recognize that *accommodation* generally describes adaptations that occur in single cells or tissues, while *acclimation* refers to adaptations that occur in physiological systems or in an entire organism. In general, cells, tissues, and sensory organs respond more quickly to stressors than do entire organisms. Despite this restricted definition of *accommodation*, distinctions between accommodation and acclimation must often be drawn artificially.

When an organism is exposed to a stressor of a constant intensity, it may respond in one of three ways. Its systems may increase, remain constant, or diminish in function with respect to time. One index of successful adaptation is a decrease in intensity over time of one or more responses to a stimulus of constant intensity. This decrease in response to a constant stimulus is known as habituation, or tolerance. Habituation to cold, heat, and low or high concentrations of oxygen are well-known forms of physiological adaptation, but not all decreases in response intensity to the stimulus of a constant stressor are adaptive. Many responses show fatigue, in which the strength of the response diminishes under the repeated or prolonged influence of the stimulus. For instance, people who experience prolonged exposures to very hot conditions may show decreases in the rate of sweating, in the past called sweat gland fatigue. Whether this is truly fatigue is arguable, but decreased sweating diminishes evaporative cooling of the skin and results in a rapid rise in body temperature that may herald the onset of heatstroke.

Although it may not be especially useful to define accommodation, acclimation, or acclimatization on the basis of time of onset or duration, two concepts about time and adaptation are quite useful. First, the most extreme conditions generally produce the fastest, greatest number of, and most intense biological responses. If the length of an extreme exposure is too brief, however, even a very intense response may not protect from a second exposure. On the other hand, if the extreme exposure lasts too long, the individual may not live long enough to adapt to the new conditions. As a general rule, gradual adaptation is the most effective way to acclimatize. Second, all types of adaptive responses are completely reversible unless they have been maintained long enough to permanently alter the structure (morphology) of a tissue or organ. However, the time course of reversal (or kinetics) and the ordering of events often differ from the events at the onset of the stress, even when permanent changes have not been introduced in the organism.

Any process that completely or partially reverses an adaptive response is known as de-acclimation or de-acclimatization. These terms refer to all biological changes that occur after an exposure. For example, shivering in the cold begins only after a small decrease in body core temperature (critical temperature), but it stops when skin temperature returns to normal, even if the body core temperature remains low. On the other hand, many other responses to cold do not return immediately to pre-exposure levels after leaving the cold. De-acclimation responses occur at different rates and without temporal relationship to their rate of onset.

These definitions of adaptation, acclimatization, acclimation, and tolerance will allow us later to examine some real episodes of human survival. The goal is to understand crucial biological factors that confer survival advantages in different environments. Close attention will be paid to the importance of physical diversity, because physical attributes may play a pivotal role in buying enough time to adapt or be rescued from a particular environment. For instance, massive obesity, a harbinger of premature death in the modern world, holds a clear advantage when it comes to surviving famine or awaiting rescue from very cold water after a shipwreck, but obesity also makes it much more difficult to acclimatize to exercise in hot weather.

These concepts provide a working vocabulary, but they also teach an important lesson about human adaptation. Whether an individual adapts to and survives extreme conditions depends on the balance between two of the four key variables: the natural environment and the biological limit of an individual with specific physical attributes. Therefore, exploring a range of conditions that involve strain will help elucidate the interplay among individual attributes and each of the stressors involved in promoting adaptation in humans. This requires determining how acclimation to a specific stressor by an individual contributes to the probability that the person will adapt successfully and live through the exposure. Before assuming the problem is quite so simple, however, we need to acknowledge the rich complexity of each environment and the important differences between acclimation to a single stressor and acclimatization to an environment.

# 3

# Cross-Acclimation

A wonderful opportunity to illustrate the rich complexity of biological responses to natural environments arises when a difference is encountered between acclimation to a single stressor in an environment and overall acclimatization. This biological complexity can be conceptualized in a scientifically useful way. The key concept is based on the hypothesis that acclimation to a single stressor triggers a general pattern of responses that could augment or interfere with acclimation to a second, independent stressor. The name given to this process is *cross-acclimation*, which is defined as the influence of earlier adaptation to one stressor on subsequent adaptation to a new environment that may or may not contain the initial stressor.

## The Complexity of Adaptation to Environment

Cross-acclimation demonstrates the importance of the interplay among different stressors that influence the integrated, or overall, response to a complex environment. To complicate the issue a little further, cross-acclimation may result in either positive or negative acclimatization to a new environment. For example, earlier adaptation to cold may help an animal survive a subsequent exposure to ionizing radiation, but it interferes with its ability to survive a lack of oxygen. In addition, lack of oxygen (hypoxia) decreases the shivering response to cold. If these principles operate in mountaineers, then it should be more difficult to ascend to the summit of Mount Everest in winter than in summer. Climbers know this intuitively from experience. The point can be illustrated by examining the history of ascent of Mount Everest.

The first successful ascent of Everest, in May 1953 by Sir Edmund Hillary and Tenzing Norgay, occurred in a season when climbing conditions were nearly optimal. Hillary and Norgay used supplemental oxygen to make their final ascent to the summit. The first ascent in winter using oxygen occurred almost a quarter of a century after the pioneering springtime ascent. The first ascent in summer without oxygen, in 1978 by Reinhold Messner and Peter Habeler, also preceded the first winter ascent without oxygen by a decade. Other factors, climatic and technical, certainly contribute to better success in summer, but the point is that the colder it is, the harder it is to climb at extreme altitude.

## Positive and Negative Cross-Acclimation

If one understands the elements of cross-acclimation, it is easier to grasp the processes of adaptation in general. As noted, both positive and negative cross-acclimations are possible. When the first experiments on cross-acclimation were published more than fifty years ago, there was no rational basis for predicting whether acclimation of a human or other mammal to a single stressor would produce a positive or negative response upon exposure to another stressor. At the time, the results of some cross-acclimation experiments raised eyebrows, but as more data became available, a coherent story was gradually pieced together from the observations. For instance, the physiological responses to cold and hypoxia share certain important similarities, such as an increase in the release of stress hormones from the adrenal gland. Taken at face value, this might be interpreted as evidence of positive cross-acclimation.

In 1937 the famous physiologist Hans Selye proposed that all types of acclimation involve activation of a general adaptation syndrome (see Selye, 1993). Selye postulated that all stresses invoke a nonspecific reaction, beginning with alarm, followed by resistance, and, if the stress is too great, culminating in exhaustion. In a more specific version of this theory, Conn and Johnston in 1944 proposed that an increase in the activity of the hypothalamic–pituitary–adrenal (HPA) axis is primarily responsible for general adaptation, which today is known as an important component of acclimatization.

To frame this hypothesis in modern physiological terms requires an appreciation of the purpose of a small and ancient phylogenic region of the brain called the hypothalamus (Swanson, 1999). The hypothalamus regulates many critical bodily functions, such as temperature, sleep, and appetite. It forms the floor of the third ventricle, one of the four spaces within the brain that is bathed in cerebrospinal fluid. A vast amount of evidence indicates that the hypothalamus regulates the expression of a range of behaviors critical to the survival of an individual as well as to a species as a whole. The hypothalamus is attached by a slender stalk to the pituitary, often called the master gland. It is also connected to numerous areas of the higher brain and even has projections to the spinal cord. Some of the neurons in the hypothalamus demonstrate unique sensitivity to temperature: some increase firing with heat, and others increase firing with cold. These responses are coordinated with spinal cord and peripheral receptors to provide integrated thermoregulatory responses, including changes in behavior.

In the hypothalamus neuroendocrine cells secrete minuscule quantities of tiny proteins, or neuropeptides, that stimulate the pituitary gland (Fig. 3.1). This chemical stimulation in turn causes cells of the pituitary to release hormones into the bloodstream. These hormones bind to specific receptors, usually on the surfaces of cells, and alter their activities. They also regulate secretion of the neuroendocrine cells in the hypothalamus to maintain control of the system.

One of the most important pituitary hormones is adrenocorticotrophic hormone, or ACTH, which is responsible for stimulating the adrenal gland to release stress-

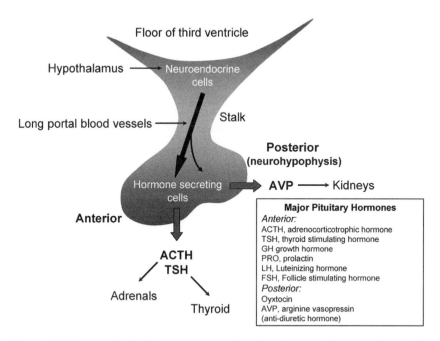

**Figure 3.1.** Neuroendocrine factors necessary for adaptation to environmental stress. The stress response requires proper functioning of the hypothalamus, pituitary gland, and related hormones. The pituitary produces many hormones, but three, adrenocorticotrophic hormone (ACTH), thyroid stimulating hormone (TSH), and arginine vasopressin (AVP), are vital to defend against environmental stress, particularly cold, heat, and dehydration.

related hormones. These stress hormones, known as corticosteroids, have been implicated in all forms of acclimation. In addition, many stressors activate the well-known "fight-or-flight" response, mediated by the sympathoadrenal system, and the release of hormones called catecholamines, such as epinephrine and norepinephrine. In addition to ACTH, the pituitary gland produces two other hormones critical to survival in extreme environments. These hormones, discussed later, are thyroid stimulating hormone (TSH) and arginine vasopressin (AVP). Although many types of acclimation are associated with hormonal responses, they are not sufficient to explain all the processes involved in acclimatization. This point is illustrated by studies of cross-acclimation.

When humans or lower animals are exposed to two independent stressors in a closely ordered sequence, information about cross-acclimation between the two stressors can be obtained. In such cross-acclimation studies a negative relationship can be demonstrated between prolonged exposure to cold and acute exposure to hypoxia. In other words, animals that adapt successfully to the cold fare less well when exposed to high altitude than do animals not adapted to cold. In the opposite situation, animals that are successfully adapted to high altitude are less tolerant of cold than are animals at sea level.

Although many plausible explanations exist for this phenomenon, it is sufficient to say for now that stress hormone secretion from the adrenal gland either serves different purposes under the two circumstances or interacts with supplemental responses to either or both conditions. As one might reasonably surmise from these observations, some evidence also supports a positive effect of acclimation to heat on tolerance to subsequent hypoxia.

Notwithstanding the negative effects of cold on tolerance to hypoxia and vice versa, some cross-acclimation studies show positive effects of cross-acclimation to one stressor on a variety of others. For example, some physical conditioning regimens appear to improve heat tolerance. Positive cross-acclimation between one stressor and other independent stressors suggests that stress in general recruits common biological mechanisms that protect against many types of environments. Thus, Conn's original stress hormone hypothesis is much too simple to explain cross-acclimation, and many novel biochemical factors have been discovered that contribute to and regulate these processes.

## Biochemical Mediators of Physiological Adaptation

In the last 50 years more than 100 different neuropeptides and hormones have been discovered that are produced by the body in various amounts and combinations under the influence of different stresses. Undoubtedly, many more of these mediators remain to be discovered. Most of these molecules have both unique and redundant functions. For instance, at least a dozen mediators are able to influence the temperature of the body. A good example is the production of thyroid hormone, which is increased by cold stress and decreased by heat stress. Thyroid hormone helps regulate heat generation (thermogenesis) by the body.

The cells of the body's immune system produce a variety of small peptide mediators called cytokines, which regulate many physiological processes and are involved in structural remodeling of organ systems. These cytokines are produced in and released from various cells and bind to specific sites, or receptors, on other cells to influence their behavior. This arrangement is part of a critical biological process known as cell-to-cell communication, or intercellular signaling. For example, some of the members of an important family of cytokines, known as interleukins, regulate body temperature during infection. In fact, the first of these mediators to be discovered, interleukin-1 (IL-1), was originally known as endogenous pyrogen because of its powerful ability to induce fever.

In the 1970s a group of proteins was discovered in the body that could be induced by heat stress. These proteins, known as heat shock proteins (Hsp), are present in only small amounts in normal cells. When living organisms, including humans, are exposed to exceptional heat stress, molecular processes are modified to minimize damage to cellular structures. The heat shock response was first observed in the fruit fly, *Drosophila*, in the 1960s when the nuclear material (chromatin) in the cells of the fly's salivary gland was found to "puff," or unwrap, after the insect was exposed to heat stress. The unwrapping of chromatin was necessary to activate genes in the nucleus and allow new proteins to be synthesized in response to the heat shock.

The term *heat shock* is too restrictive to describe the function of these proteins, which respond to many other cellular stresses, including but not limited to inflammation and oxidative stress. In addition, many of these proteins play essential roles during unstressed conditions. Today the term *stress proteins* is preferred, but the original observations on heat tolerance launched the studies that led to understanding the principles of induction of protective Hsp genes. Thus, the heat shock response has become associated with universal defense against noxious external stimuli.

## Stress Proteins and the Stress Response

Stress proteins are found in all cellular compartments and range in size from 10 to 150 kilodaltons. By convention stress proteins are named by size (e.g., Hsp70, Hsp90, Hsp100; the corresponding genes are designated as *hsp*70, etc). A key property of Hsp70 and Hsp90 is their ability to bind newly synthesized or damaged polypeptides and restore native structure to the latter. All Hsps address problems in protein folding, and many affect protein synthesis or modification.

Stress protein expression is induced by activation of transcription and selective protein translation (Fig. 3.2). Transcription, the process of synthesis of mes-

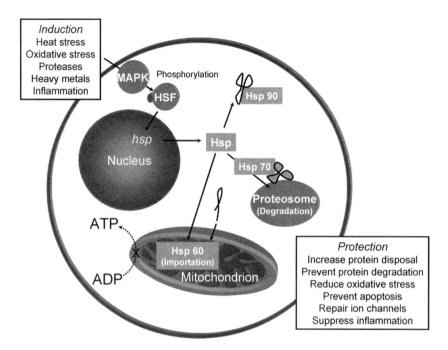

**Figure 3.2.** Induction of the cellular stress response. Exposure to heat and a number of other types of stress (box, upper left) invokes a common cellular response that stimulates synthesis of heat shock proteins at the expense of other protein synthesis. The stress response protects the cell by multiple mechanisms, some of which are indicated in the box at the lower right.

senger RNA from DNA (genes) in the nucleus, and translation, the process of reading a message and translating it into polypeptides, are fundamental life processes. Despite a great deal of research, precisely how stress stimulates these processes is unknown, but once triggered, a set of transcription or heat shock factors (HSFs) activates *Hsp* genes. HSFs regulate *Hsp* genes not only during adaptation but during normal growth and development. In primitive organisms such as yeast, one gene encodes HSF, whereas as many as four HSFs are present in plants and animals, including humans. HSF1 undergoes phosphorylation by kinase enzymes, in particular members of a family of mitogen-activated protein kinases (MAPK), during physical stress. Phosphorylation allows HSF1 to enter the nucleus and bind to specific sites in genomic DNA called promoters, which enable transcription. During unstressed conditions, the DNA-binding and transcriptional activity of HSF1 are inhibited. Studies have demonstrated an important role for energy depletion in activating HSF1-DNA binding. In disease states HSF1 is activated by factors that increase protein damage. However, activation of HSF1 does not require new protein synthesis because HSF1 is present but inactive in the unstressed state.

One particularly interesting aspect of the stress response is that it presents cellular synthesis of most new proteins except for Hsp. When genes that direct Hsp production are activated, cells produce these and only these proteins in meaningful amounts for some time. These proteins at high concentrations protect cells against a host of physical and chemical insults by several mechanisms. Some Hsp defenses reduce stress by scavenging free radicals or eliminating other harmful metabolic products of stress. Other stress proteins interact with cellular receptors for hormones, such as thyroid and steroid hormones, and restrict their activities. After the stress subsides, the receptors are liberated, and hormone effects again occur freely.

Proteins of the Hsp70 family are the major stress protein in eukaryotic cells and are among the most abundant systems for protecting against cellular injury. Hsp70 are well known because they serve as molecular "chaperones." Chaperone proteins guide the correct folding of new proteins as they are released from the cell's synthetic machinery, the ribosome. The ribosome reads the code in the messenger RNA exported from the nucleus and assembles individual amino acids into chains, or polypeptides (translation), that constitute the primary structure of the protein. Primary structure is not sufficient to complete a functional protein; the polypeptide must be folded in space. Folding imparts the proper globular shape, or secondary and tertiary structure, to proteins. In addition, many proteins undergo complex posttranslational assembly and/or modification, such as chemical addition of glucose molecules (glycosylation). These structural modifications allow the molecular interactions necessary for proper function, for example, as enzymes.

Chaperones also associate with mature proteins to modify their activities, cellular locations, and metabolic fates. For example, some chaperone proteins facilitate unfolding and refolding of other proteins as they move around the cell. Chaperones can also pass the protein molecule to other chaperones as it enters different compartments or organelles within the cell.

Improper unfolding and other types of irreversible damage to proteins may lead to loss of function or nonspecific aggregation of proteins. Such modifications, known as denaturation, are associated with exposure to excessive heat. Denaturation of proteins causes changes in the physical state of a raw egg when it is boiled. The presence of chaperones protects heat-susceptible proteins in the cell from denaturation and allows the molecules longer use times. Another group of Hsp, the ubiquitin proteins, facilitates orderly disposal of proteins when denaturation goes too far. Denatured proteins are "ubiquinated" and escorted to a disposal site, the proteosome, where they are taken apart and their component amino acids recycled for incorporation into new proteins.

The stress response is an important form of molecular adaptation. It focuses homeostatic mechanisms on cell survival by preventing too many proteins from being damaged by stress. As long as Hsp are expressed, cells deal more successfully with subsequent heat stress as well as many other stresses. Understanding adaptation to environment at the molecular level is still in its infancy, but herein lies the secret of cross-acclimation. Like the rest of biology, adaptation ultimately must be understood in terms of these molecular processes because they underlie every long-term physiological and disease response. Moreover, molecular adaptation takes time to develop and must be linked appropriately to the integrated physiological responses that protect the body at the onset of stress.

There will be more to say on the effects of Hsp, cytokines, neuropeptides, and hormones on adaptation in the following chapters. For now, it is enough to recognize that cross-acclimation has a sound biological basis that begins with molecular interactions at the level of single cells. Before calling on these molecular responses to do their jobs in a changing environment, the body must have adequate supplies of food and water. Even if lack of food and water are insufficient to cause death, malnutrition and dehydration diminish acute and chronic human tolerance to every conceivable stressor. In particular, dehydration impairs tolerance to heat and cold, and malnutrition to cold and disease.

# Food for Thought

In 5000 years of recorded history, nearly 500 major famines have been documented in various parts of the world. Even in the twenty-first century, about 1 million people starve to death worldwide every year. When one realizes how difficult it is to starve to death today, the magnitude of this tragedy becomes incomprehensible. The contrary view that almost 6 billion people are fed on the planet every day reminds us of the technological tour de force that is modern agriculture. Fewer than 1 in 6000 people starve to death each year in today's world. However, the true consequences of human starvation and malnutrition are far larger and more insidious than is the body count. These include epidemics of related diseases, lost productivity, and loss of tolerance to environmental stress, particularly to heat and cold.

## A Brief Overview of Human Starvation

The human impact of starvation has been an area of controversy among scholars, physicians, clerics, and politicians for hundreds of years. In 1798 the economist T. Robert Malthus (1766–1834) wrote in his famous "An Essay on the Principle of Population" that unchecked population growth would result in "gigantic inevitable famine" leading to the extinction of civilization. Within a few years, however, Malthus realized that this prediction lacked a basis in evidence, and his later writing on population was not so dire. Modern demographers also discount the idea that famine has been an important constraint on human population growth. Rapid population growth and overcrowding obviously produce problems, but the worldwide famine Malthus feared has failed to materialize. Nonetheless, this theory is occasionally resurrected to explain famine, poverty, and disease. For instance, administrators in the British Raj of the nineteenth century argued against famine relief as expensive and futile meddling with the natural order of the Indian subcontinent. As this example poignantly illustrates, population growth is often made the scapegoat for intractable poverty, shortage of arable land, and unbridled epidemics with no regard to human biology and technology.

In the course of the twentieth century, despite continuous expansion of the population, critical food shortages became less frequent and less severe except in sub-Saharan Africa, where there have been dreadful famines in Ethiopia, Somalia, and the Sudan the last twenty-five years. To appreciate the reasons for this, consider the elements of famine prevention. In geopolitical terms, there are at least three main elements of prevention. The first is available surplus agricultural production beyond that needed to sustain the population. The second is an effective transportation system within the region and well-developed connections to the rest of the world. The third is a democracy or other form of government that is egalitarian or endowed with a social conscience.

It is not my intent to argue the merits of these elements of famine prevention, which allow the following points. Generally, agricultural surplus makes food shortages less frequent, and those that do occur tend to be less severe or persistent. Efficient transportation makes it possible to deal more effectively with food shortages, and egalitarianism brings about political pressure to ensure action is taken against starvation. In a democracy, the press usually raises the alarm, particularly in remote regions, and public opinion demands action by the society to alleviate the food crisis. This view of famine prevention also makes it apparent why the major famines of the last twenty-five years were precipitated by war, which effectively disrupts all elements of famine prevention.

## Starvation: An Affliction of the Very Young and the Very Old

For centuries it has been known that the very young and the very old are the first to die in famine. A great deal of attention is paid to starvation of children because of their small size and special nutritional requirements for growth and development. In children, growth retardation, or stunting, has often been used to estimate the magnitude and extent of nutritional problems in different parts of the world.

The severity of stunting is often assessed on charts or nomograms of height compared to age. Such charts are only approximations, but the approach has been employed for many years by the World Health Organization (WHO) as a global indicator of the well-being of children. Children who suffer growth retardation also show greater susceptibility to infectious diseases such as diarrhea, malaria, and pneumonia. Of the children under age five who died in the Ethiopian famine of 2000, death was caused by malnutrition in only 23%; respiratory infections (6%), measles (22%), and infectious diarrhea (37%) accounted for most of the rest.

As stunting becomes more severe, it produces three major effects. It delays mental development, decreases intellectual performance, and increases childhood mortality. Those who live into adulthood have a lower work capacity and a lower level of socioeconomic success. The biological lesson is that for the developing central nervous system and brain to function well, specific nutrients are required in relatively large amounts. In adults brain metabolism uses 20% of the body's fuel, but in children the developing brain consumes almost half the daily energy supply. Most of this substrate is glucose, although during starvation the brain can also oxidize ketones. How well the alternative substrate serves a developing child's brain is unknown. In addition, childhood growth retardation is a good indicator of maternal nutrition because stunting begins in utero, and poor intrauterine growth predicts poor postnatal growth.

There is some good news about childhood malnutrition in developing countries (de Oris et al., 2000). Growth retardation measured by stunting fell progressively worldwide from 47% in 1980 to 33% in 2000. The bad news is that a third of all children in developing countries are stunted by the age of five. Two-thirds of these children live in Asia, mainly in South Central Asia, and one quarter of them live in Africa. Better understanding of this problem and meaningful investment in policies to improve the physical growth and mental development of children are needed to prevent detrimental effects of malnutrition throughout the human life cycle.

## Assessing the Severity of Starvation

Many attempts have been made to assess the severity of starvation, but short of starving to death there are no universally accepted criteria for it. The time-honored techniques are designed to quantify the extent of body wasting (Cahill, 1970). In children, these include mid-upper-arm circumference (MUAC) and a weight-for-height index. A weight-for-height index of less than 80% is considered severe wasting, and less than 70% is considered critical wasting. In adults the amount of body swelling (hunger or famine edema) and body mass index, or BMI, are often used to predict risk of death. The severity of swelling appears to be a better predictor of death in men than in women for reasons that are not entirely clear. Men also tolerate starvation less well than do women for reasons that are also unclear. This is evidence that women are hardier than are men.

BMI is the body weight in kilograms divided by the height in meters squared ($m^2$). For an average 70-kilogram man of 1.8 meter height, the BMI is roughly 22 (normal is $\geq 19$). The lowest BMI compatible with life is approximately 10, or for an average man 1.8 meter tall, 32.4 kilograms (roughly 71 pounds). Because such severely emaciated people are usually unable to walk or even stand, cannot regulate body temperature, and often have pulmonary or intestinal infections, expert medical care is their only chance for survival (Collins, 1996). A low BMI also correlates with a greater risk of dying from starvation despite refeeding, although extreme emaciation can be treated successfully. Individuals with BMIs of less than 9 have survived famine provided they received care in specialized nutrition centers. It is also important to note that swelling falsely increases BMI because of the weight added by retention of fluid.

Adults can survive for weeks or months without food depending on the amount of fat in the body. A 70-kilogram man can fast for about seventy days, losing of all but 3% of body weight in fat and one-third of lean body mass. A small amount of fat is essential to maintain the functions of the brain, bone marrow, and cell membranes. In a healthy, nonobese adult, assuming adequate salt and water intake, death from starvation can occur virtually any time after 50% of the body mass has been lost.

In the Dutch famine of World War II, previously well-nourished victims suffered a year of hunger followed by six months of severe starvation compounded by the stress of war and foreign occupation. Those whose weights fell below 50% of normal almost always died. Infection also played a key role in mortality, particularly pneumonia. Women in early pregnancy aborted, and those in the last trimester gave birth to underweight babies.

The costs of malnutrition are not limited to the immediate victims of famine; the effects also occur in other ways (Huxley et al., 2000). In 1948 a high incidence of birth defects was reported in babies born to mothers who had survived the severe hunger and psychological stress of German concentration camps in World War II. Among 1430 babies born to these women at the Klebanow obstetrics clinic, 4% had major birth defects, whereas in the same clinic only 1% of babies born to women who had not undergone such stress had birth defects. Among the 1430 babies of Holocaust survivors, twelve had Down's syndrome, a prevalence rate of 0.8%, or fifty times the expected prevalence of the syndrome in the clinic at the time.

Over the last fifty years it has become apparent just how critical maternal nutrition is for avoiding fetal growth retardation and birth defects of the central nervous system. If maternal nutrition is inadequate for normal fetal brain development, a child is more likely to be at a significant intellectual disadvantage in life. An example is inadequate maternal intake of just a single water-soluble vitamin, folic acid.

Folic acid is necessary to synthesize nucleic acids, both DNA and RNA. Infants of mothers who receive inadequate dietary folic acid have low birth weights and high incidences of growth retardation and neural tube defects. Normal development of the placenta, embryo, and fetus requires rapid cell division, which depends on folic acid. During gestation dietary folic acid can prevent neural tube defects, particularly in mothers who have previously had a child with a defect. Relative folic acid deficiency can also occur because even minor genetic defects can increase its metabolic utilization. Thus, dietary deficiencies that have little obvious maternal effects can manifest themselves in the child because the fetus has a higher requirement for the nutrients of growth and development.

## Why Children Die of Starvation

The marked susceptibility of children to starvation has been demonstrated in famine after famine. A young child can live for only about six weeks without food. In the great civil war famine of 1992 in Somalia, mortality statistics from more than a dozen epidemiological studies, despite significant methodological differences and statistical problems, showed the same trends for excess mortality in children under age five. Although the death tolls cited for all major famines have large margins of error, a threshold of 1 death per 10,000 people per day defines a humanitarian emergency. In Somalia in 1992, death rate estimates in adults were 7.3 to 23.4 per 10,000 people per day, while in children under age five they were 16.4 to 81.0 per 10,000 per day. At the famine epicenter in the regional capital of Baidoa, it is estimated that up to 75% of displaced children under age five had died by November 1992.

To understand childhood starvation, one must understand how the normal human body stores fuel (Bines, 1999). These fuels, known as macronutrients, are stored in three ways. In the body of a normal 70-kilogram adult, about 0.5 kilogram is stored as complex carbohydrate (glycogen), 6 kilograms as protein, and 15 kilograms as fat. During a day's fast energy is provided by the glycogen in the liver and by converting protein to glucose (gluconeogenesis). During prolonged fasting body fat provides energy (150 grams/day) in the form of ketones from fatty acids. Body fat and muscle protein (20 grams/day) are used in roughly equal amounts to generate glucose. During starvation energy use decreases to conserve the body's macronutrient reserve. Although individuals of normal body weight can fast for about seventy days, highly obese individuals can fast for more than a year depending on the amount of fat in reserve. On the other hand, the smaller bodies and higher metabolic rates of children cause them to deplete their macronutrients more quickly.

To appreciate this problem in detail requires working definitions of human starvation (Golden, 2002). Fasting refers to the absence of food intake, and starvation is the physiological result of fasting, or insufficient macronutrient intake, for a prolonged period. Malnutrition refers to long-standing deficiencies of either macronutrients or micronutrients, such as vitamins and minerals. Malnutrition can be due to an inadequate nutrient supply in the diet or diseases that interfere with the absorption of nutrients from the alimentary tract. Protein–calorie malnutrition (PCM) refers to inadequate intake of protein or energy for a prolonged time. PCM often results from starvation, but starvation leads to PCM only after the body's energy reserve has been depleted. PCM occurs in two overlapping forms, known as marasmus (primarily calorie deficiency) and kwashiorkor (protein deficiency).

The grave sign of starvation, edema, is associated with a diet chronically deficient in protein. Low-protein, or famine, edema may be exacerbated by thiamine deficiency (or beriberi), which causes a form of heart failure that interferes with effective distribution of blood to the organs. Severe protein deficiency also leads to failure to synthesize plasma proteins, such as albumin, which causes the volume of circulating blood plasma to fall. Because plasma proteins regulate the osmotic (colloid) pressure in blood relative to tissue, decreases in plasma protein concentration to about half normal values cause tissues to retain edema fluid. Although protein deficiency has long been implicated in the etiology of starvation edema, it is certainly not the only factor. Starvation is associated with the depletion of essential antioxidant compounds, particularly glutathione, and the accumulation of free iron, which make tissues susceptible to oxidative damage. However, the link between oxidative stress and edema is not yet clear. In addition, starvation edema is worsened by energy deficiency and the depletion of cellular reserves of potassium. When famine edema is prominent, the skin pits, or dents, when pressed briefly by a finger. The appearance of severe pitting increases the risk of dying from starvation by as much as five to ten times.

Famine edema may actually worsen when starving people are suddenly provided adequate nutrition. The extra food increases the body's amount of salt, which is retained with water. This helps produce the so-called edema of refeeding. Refeeding edema may worsen because the sugars in food cause more insulin to be released from the pancreas, which, in addition to lowering blood sugar, causes the kidneys to reabsorb extra sodium. Refeeding edema also can be worsened by a diet high in protein. Under these conditions patients may die suddenly for reasons that are not yet understood fully. Indeed, many survivors of Nazi concentration camps died when given free access to food. This taught medical science about the dangers of refeeding and fostered measures to avoid untoward nutritional recovery syndromes.

## Other Critical Factors in Human Starvation

The decisive manifestations of starvation depend not only on age but previous nutritional state, duration and severity of food deprivation, and pre-existing disease. Morbidly obese adults who voluntarily fast on water, vitamins, and potassium salts for many months lose their hunger after a few days. Apart from feeling tense, they have few symptoms. Physical strength gradually decreases, and after some weeks tachycardia and positional decreases in blood pressure (orthostasis) become apparent. Plasma electrolytes, lipids, proteins, and amino acids are unchanged because the body subsists on its energy store, including amino acids from protein breakdown. In contrast to starved obese subjects, nonobese individuals often fixate on food and suffer from severe, prolonged, and distracting hunger pains.

Starvation in association with disease or environmental stress is particularly deadly. Because starvation compromises immunity, microbial factors come into play; infectious diarrhea and pneumonia have been mentioned, but in some parts of the world, the infections of opportunity are malaria, tuberculosis, and AIDS. In addition, heat and cold take their toll because starving people do not adapt as well to extreme climatic changes. Cold is particularly devastating because the body's major defense, shivering, consumes huge amounts of energy.

The body's expenditure of energy stores (fat, glycogen, and protein) during starvation is different than it is during other types of strain or stress. The stresses of physical trauma and acute illness such as infection affect metabolism. However, instead of depressing metabolism, conserving protein, and using body fat for energy, acute stress accelerates metabolism. Energy expenditures are met primarily by breaking down muscle protein, called catabolism. Skeletal muscle protein is broken down to provide amino acids to synthesize glucose. Muscle catabolism is facilitated by stress hormones, including catecholamines, cortisol, and glucagon, that are released at high levels into the blood. The stress response also causes cytokines to be released, including tumor necrosis factor (TNF) and interleukins, such as IL-1 and IL-6. TNF was originally called cachexin because it was first discovered as a cause of wasting (cachexia) in patients with advanced cancer. TNF and other cytokines bind to receptors in target cells and produce a wide range of molecular responses that require energy use. In severe infection the amino acid glutamine, which is released from muscle protein, is a critical substrate for the functioning of cells of the immune system. Protein catabolism during infection can approach 300 grams per day, which will waste half the muscles of an adult in about six weeks.

Malnourished people commonly suffer injuries or infections that impose cytokine-mediated demands on the body. If the protein–calorie malnutrition is unchecked, it leads within days to heart and kidney failure, worsening edema, and atrophy of intestinal villi, microscopic fingers that protrude into the lumen of the gut to increase the surface area for absorption of nutrients. Malnourished patients also deplete their stores of intracellular potassium and essential minerals such as phosphorus and zinc. The catabolic state diminishes the ability of immune cells to function, further increasing the risk of infection and eventually leading to death.

In most famines, with a few noteworthy exceptions, severe vitamin and mineral deficiencies have not been at the forefront of clinical problems. Although vitamins and minerals are dietary essentials, the amounts of minerals and vitamins available during starvation are often about equal to or slightly less than those needed to serve the shrinking body. However, in modern famines in Ireland, Ethiopia, Somalia, and the Sudan, the appearance of scurvy was a significant exception to this generalization. In recent Asian and African famines, the ravages of vitamin A deficiency on the eyes, causing night blindness, have been common. Depletion of intracellular potassium has already been mentioned; similar losses occur for other essential minerals, such as iron and copper, particularly when chronic or recurrent diarrhea develops.

When nutritional needs are increased on the backdrop of hunger, as in infection, vitamin and mineral deficiencies become manifest. Vitamin deficiency syndromes, such as beriberi, pellagra, and scurvy, emerge, and people die suddenly. Well-meaning but poorly conceived attempts to feed the starving can result in death for similar reasons. Increased protein intake increases requirements for potassium, zinc, magnesium, vitamin A, and water-soluble B vitamins such as riboflavin and niacin. The risk of inducing severe vitamin A deficiency by overstimulating protein intake is so great that a leading relief agency has cautioned that dry skimmed milk, a vital food staple, is likely to be deficient in vitamins A and D unless it comes from UNICEF, Canada, or the United States. High caloric intake also creates an increased need for thiamine and other B vitamins. In the course of famines, relief workers have encountered deficiencies of virtually all essential minerals and vitamins brought on by diets that were incomplete in the face of the protein and calories needed to restore body weight and composition.

## Starvation and Obesity: Strange Bedfellows

In one respect, starvation stands in stark opposition to human obesity, which creates a unique set of medical and psychosocial problems. In another respect, the two are linked inseparably. Many ecological studies have shown a U-shaped relationship between body weight and mortality, with individuals at very low and very high weights at increased risk (Fig. 4.1). This relationship has been found even after adjusting for confounding influences, such as smoking and preexisting disease. The major causes of death between the two extremes of the curve are different, and in some studies the curve is J-shaped (flatter at low body weight), with the greatest increase in mortality rate associated with BMIs of 30 or more. Men with BMIs of 35 or more have nearly three times the relative risk of death, mostly from cardiovascular disease, compared to individuals of BMIs of 20. The shape of the higher end of the curve suggests that greater body weight is associated with greater strain on one or more critical body functions, such as the cardiovascular system.

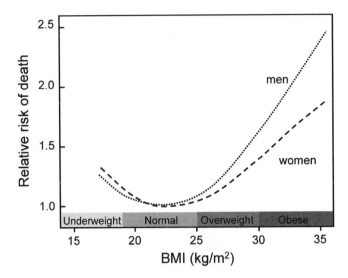

**Figure 4.1.** Body mass index and relative risk of death. The U-shaped curves indicate increased mortality in both overweight and underweight individuals of both sexes. Data adapted from the National Heart, Lung and Blood Institute (National Heart, Lung and Blood Institute. 1998. *Clinical Guidelines on the Identification, Evaluation, and Treatment of Overweight and Obesity in Adults: The Evidence Report.* Bethesda, Md.: National Institutes of Health).

Obesity is part of the "deadly quartet" of linked metabolic risk factors (obesity, diabetes, hypertension, and hyperlipidemia) for cardiovascular disease. The connection between obesity and cardiovascular disease is clouded by some controversy because obesity is also associated with sedentary behavior and a less healthful diet, which independently affect cardiovascular health. It is also linked to stroke, sleep-disordered breathing (apnea), liver and gall bladder disease, venous thrombosis, arthritis, and certain kinds of cancer.

The definitions of overweight and obesity are an area of confusion. The BMI is perhaps the best single measure of obesity (better than weight-to-height tables), but it does not measure body fatness or fat distribution, which is an independent predictor of health risk. Because body fat is not measured, muscular persons such as athletes may be misclassified as obese using BMI alone. Therefore, BMI and body fat distribution are often used together to evaluate the health risk of obesity. The National Institutes of Health 1998 Guidelines on Obesity in Adults define overweight as a BMI of 25 to 29.9 and obesity as a BMI of at least 30. Defining overweight as a BMI of at least 25 is consistent with recommendations of the World Health Organization. Defining obesity as a BMI of 30 or more places roughly one-quarter of women and one-fifth of men in the United States in this category.

A key predictor of health risk in obesity is the distribution of body fat. Body fat may be located in the abdomen (android pattern) or at the hips and thighs (gynoid pattern). The android pattern often reflects accumulation of visceral fat around abdominal organs and is associated with metabolic derangements, including hypertension, diabetes, and hyperlipidemia. Thus, even for two people at the same weight and height, an individual with more visceral fat is more likely to have or develop health problems from obesity. Methods to assess visceral fat accurately are not available for routine screening, but a reasonable surrogate measure is waist circumference. A waist circumference of at least 35 inches (88 centimeters) in women or 40 inches (102 centimeters) in men is associated with increased health risk. Although these guidelines provide reasonable risk estimates for the general population, there is significant individual variability in the amount of visceral fat at a particular waist circumference. Despite the mounting evidence that obesity is deleterious to health and longevity, it has reached epidemic proportions in American society.

## The Molecular Basis of Obesity and Hunger

Obesity is not just a matter of voluntary over-eating; it has a molecular basis in metabolism, hunger, and satiety. Over the past decade dazzling progress has been made in the molecular understanding of the cause of obesity. Discovery of an obesity gene called leptin, after the Greek *leptos*, or "thin," in mice has revolutionized the way scientists think about diet and nutrition. The leptin gene is one of several identified as potential causes of obesity.

The fascinating history of leptin began with parabiosis experiments in which the blood circulations of two living animals were connected. Such experiments provided the first evidence for the existence of a circulating satiety factor. When the circulation of a normal animal is paired with that of an animal made obese by injury to a region of the brain called the ventral medial hypothalamus (VMH), the normal animal stops eating because its appetite becomes suppressed. In mice with hereditary obesity, called ob/ob mice, weight gain can be suppressed by parabiosis with normal mice. This finding suggested to scientists that a satiety factor was missing in ob/ob mice.

In 1994 J. M. Friedman and colleagues at Rockefeller University successfully mapped and cloned a key obesity gene. The *obese* gene encodes a 167-amino acid protein called leptin, which is expressed exclusively in fat cells, or adipocytes (Friedman and Halaas, 1998). A homozygous mutation in its DNA sequence confirmed it as the cause of obesity in ob/ob mice. In addition to obesity, the ob/ob mouse exhibits other defects, including increased appetite, diabetes, hypothermia, and infertility. Each abnormality can be corrected by the administration of leptin.

These mouse studies stimulated a spate of experiments on leptin physiology to determine if a mutation in the *obese* gene was a cause of human obesity. It turned out that *obese* gene mutations can rarely be implicated as a cause of human obesity. However, the research did find a surprising correlation between body fat and leptin levels; blood leptin is high in most fat people. Leptin is also higher in women than in men, suggesting a link between estrogen or progesterone and leptin. Other steroid hormones as well as insulin increase blood leptin levels, and release of insulin and leptin is closely correlated. It is not yet known whether this observation reflects obesity or whether these two hormones regulate the activities of each other (Jequir, 2002).

A high level of leptin in obese people was puzzling because leptin suppresses the appetite. A similar phenomenon is sometimes encountered in adult-onset diabetes, when blood sugar is elevated in the presence of what should be an adequate insulin level. This is known as insulin resistance. Therefore, the finding of elevated leptin levels in obese people has been interpreted to indicate resistance to the action of leptin. A model for this, based on the concept of a "lipostat," is analogous to a thermostat by which the level of leptin in the circulation is detected by the brain (e.g., in the VMH). When the leptin content of the blood increases, heat is generated (thermogenesis) and satiety induced. In starvation fat is lost and leptin falls, leading to hunger and less energy expenditure. When resistance is present, a rise in blood leptin does not produce the appropriate responses of satiety and heat production.

The causes of leptin resistance are beginning to be worked out. A molecular receptor of the cytokine family has been identified for the leptin molecule. Mutations of the leptin receptor have been found in genetically obese animals in which the effects of leptin are diminished. Leptin receptor mutations have also been reported in obese humans who lack normal pubertal development. There are several variants of the receptor, and, unlike leptin, it is expressed throughout the body. Like other cytokine receptors, leptin receptors activate cell-signaling pathways. Thus, abnormalities in postreceptor signaling may explain some types of leptin resistance. Administration of leptin to mice causes weight loss and induces thermogenesis, which is most effective in leptin-deficient animals. Leptin administration has not been studied thoroughly in obese people, but, unless the receptor or its downstream mechanisms are badly damaged, it could become a therapy for satiety and weight loss despite resistance if sufficiently high doses can be given.

One of the more striking effects of leptin involves reproduction. Leptin administration corrects infertility in genetically obese mice. In addition, leptin advances the timing of puberty in rats, and it may block inhibitory effects of estrogen on the hypothalamus. These observations have led to the proposition that leptin is a metabolic signal that regulates reproduction in mammals. The importance of leptin in human reproductive physiology, however, remains unknown. It is intriguing to speculate that the human reproductive cycle may be inhibited during starvation by suppression of leptin. Leptin levels should be reduced by the loss of adipocytes and by hunger. Indeed, it has long been known that fertility is decreased in malnourished women. Such a mechanism could provide an effective molecular brake on the growth of human populations during times of food shortages.

Another fascinating piece of this puzzle came from the discovery in 1998 of the orexins (orexin-A and orexin-B), also known as hypocretins (1 and 2). These novel neuropeptides are produced by neurons in the lateral hypothalamic area (LHA) and the posterior hypothalamus. The LHA is a feeding center that receives signals from the liver and intestines and is involved in sensing glucose concentration in the body (Mignot, 2001). Infusion of orexins into the brain stimulates feeding in rats and mice. Thus, orexin-responsive neurons may be involved in the hunger of low blood sugar.

When the gene for hypocretin is deleted in mice, they develop a sleep disorder similar to human narcolepsy (Siegel et al., 2001). This rare disease is characterized by an inability to stay awake at appropriate times. In humans with narcolepsy, hypocretin-1 is virtually undetectable in the cerebrospinal fluid. Although the physiological role of feeding would seem, at first glance, to be independent of sleep, it is notable that orexins are involved in regulating both appetite and alertness. Stimulation of hypocretin-containing neurons from an area of the brain called the suprachiasmatic nucleus (SCN), which is involved in diurnal rhythms, may also explain some aspects of clock-dependent alertness and feeding behavior. Whether orexins are involved in the lethargy and apparent lack of hunger in advanced starvation is currently unknown (Sakurai, 2002). However, the discovery of a neurochemical switch for eating and sleeping makes evolutionary sense for imparting a survival advantage to the individual.

What does this physiology say, if anything, about the prospects for long-term survival of the human species? Will the world be able to continue to increase its agricultural capacity to meet the nutritional needs of an ever-expanding population? What will limit agricultural production? The answers to these questions are not apparent from examining individual nutritional needs, but food for thought can be gleaned from the history of famines.

The clearest message is that starvation alone is very unlikely to bring an end to humanity. Famines tend to be intermittent, regional problems, and local populations have sufficient food reserves to provide time for people to adjust reasonably well to most shortages. Food technology, rapid means of transportation, democracy, and compassion have supported and will continue to carry society a long way. However, Earth's larder is not limitless, and despite the bright future of genetic food engineering, the disparity between the food science resources of industrialized and undeveloped economies is a serious concern. Climatic cycles are also important, not only for agricultural efficiency but because heat and cold exacerbate starvation. Global warming, drought, and progressive desertification of equatorial lands are serious concerns. Finally, the company famine keeps—war and disease—is of the very worst kind. As throughout all of recorded history, this is how starvation is likely to wreak its greatest havoc on the world's population in the future.

# 5

# Water and Salt

Water is critical to the ability of plants and animals to acclimatize and survive in extreme environments, whether on the seas, in deserts, or in the high mountains. Dehydration is the most important cause of death from exposure to heat and may contribute to death during other types of extreme exposures. Lack of potable water is much more rapidly dangerous to human survival than lack of food. Furthermore, the popular notion that the human body acclimatizes to water deprivation is unfounded biologically. This misconception, or some twist on it, has caused the deaths of many a hardy but poorly informed explorer.

The misunderstanding about water stems from the fact that the human body adjusts to low water intake by conserving water. Although the body curtails its loss of water and salt in the first few days of exposure to hot climates, no scientific evidence has shown that minimum daily water requirements decrease with acclimatization. A hallmark of heat adaptation is actually an increase in the rate of water loss from sweating. Indeed, the old adage about conserving water by seeking shade in the day and walking only at night in the desert has much to recommend it. Nothing on Earth substitutes for potable water, and without an adequate supply death is inevitable. This chapter discusses how much water is enough under different conditions. It will become evident that obtaining an estimate of sufficient water is easier said than done, because requirements vary not just with environment, but also with behavior.

## The Composition of Body Water

For more than 200 years physiologists have studied the amount, distribution, and composition of body water, particularly with respect to changes in salt content. Water is the stuff of life and the medium of the *milieu intérieur*, and the human body is 50% to 70% water by mass. It is the solvent of most organic and inorganic compounds, which are transported throughout the body. At one time in the history of physiology, the composition of body water was believed to recapitulate the conditions of the primordial seas in which multicellular organisms evolved. However, this attractive teleology is untrue for reasons related to the far-ranging homeostatic capabilities of higher animals.

Despite all the research in this area, normal values do not exist for the salt and water content of the human body. The total amount of salt and water stored in the body varies so greatly among individuals that it is impossible to define normalcy. The experimental tools used to make these measurements also have built-in errors that are difficult to control. Allowances must be made for age, sex, body fat, nutritional state, dietary preference and customs, and environmental conditions.

By simply observing how much people drink, one may get a sense of this problem. Daily fluid intake may vary from as little as 1 liter (4 cups per day) for sedentary octogenarians living in climate-controlled nursing homes to more than 10 liters (3 gallons) per day for African camel drivers crossing the Sahara. The daily intake of salt (mostly sodium chloride) varies almost as much, ranging from as little as 3 grams per day for natives of Indonesia to as much as 30 grams per day for residents of Western Europe and North America. Part of this variability in salt intake is due to the natural human appetite for salt, which is shared by most mammals (Ladell, 1965).

Despite the extraordinary range of salt and water intakes found across the world, very precise ranges exist for the normal concentrations of electrolytes in the blood. For example, the plasma concentration of sodium in healthy people is regulated closely between 135 and 145 milliequivalents (mEq) per liter, and the concentration of potassium between 4 and 5 mEq per liter. If the concentration of these electrolytes falls very far outside these limits, the body suffers dire consequences. For instance, when sodium concentration reaches 120 mEq per liter, the person becomes disoriented or confused, and convulsions may occur. When potassium concentration rises above 6 mEq per liter, life-threatening cardiac rhythm disturbances may occur. Although the range of daily intake for salt and water may vary by tenfold or twentyfold, the concentration of salt in the blood is regulated to within 20% or so.

The major point is that the body behaves like a flexible reservoir for water and salt. It overflows when either one exceeds the capacity of the system and conserves when the reservoir is low. This requires very sophisticated regulatory mechanisms for both water and salt, which are demonstrated by the tight control of plasma electrolytes (Mack and Nadel, 1996).

It is also theoretically possible to consume too much salt and water, but salt and water overload is usually encountered only in patients with heart, kidney, or liver failure. Rare instances of "water intoxication" have been reported in normal people, but this requires the voluntary ingestion of unthinkable amounts of water, usually during recovery from dehydration. Of greater practical importance is the problem of water deprivation, which results in dehydration and its attendant salt imbalance in the body.

Dehydration is not just a problem for people in hot weather, but also in elderly and medically debilitated patients, who tend to have weak thirst mechanisms, and in young children suffering fever and infection. Children are more susceptible to dehydration because their plasma volume is low relative to the body surface area. Small size, once again, is a disadvantage. Dehydration is a particularly important cause of death in children in underdeveloped countries, where infectious diarrhea, such as cholera, is common. The relative loss of plasma volume from diarrhea is more rapid in small children than it is in larger individuals. Those so afflicted are less likely to receive aggressive and appropriate rehydration in time to avert death.

## Why Do Human Food and Water Requirements Differ?

Although the human body tolerates a wide range of salt and water intake, minimum requirements do exist, particularly for water. Humans can survive for weeks without food, but only a matter of a few days without water. Some observers have used a 100-hour rule of thumb for human tolerance to lack of drinking water. This number, although easy to remember, is less useful than may appear at first glance. To understand why, it is valuable to review three reasons for the more pressing need for water than for food.

The first reason why lack of water is more critical than lack of food is the problem of metabolic waste. Although small amounts of water are produced as a normal by-product of energy metabolism, other biochemical waste products, such as urea and bilirubin, are produced faster than is metabolic water. The formation of these waste products continues unabated even in the absence of food because of the breakdown of stores of carbohydrates, fat, and protein needed to support energy production. These products are excreted in the urine and bile, which requires water. Most of this water is lost from the kidneys. The intestines play a minimal role in water loss unless diarrhea is present. Hence, the amount of water that must be "wasted" due to metabolic activities depends almost entirely on the concentrating power of the kidneys.

In the animal kingdom, the human kidney is a rather modest organ. It can produce urine only two thirds as concentrated as seawater, while the kidney of the world champion sand rat concentrates its urine to five times that of seawater. For a person on a low-salt diet, this translates to a minimum obligatory urine flow of about 0.5 to 0.6 liters per day.

The second problem with water is the presence of "leaks" in the reservoir. Not surprisingly, major leaks in the system occur because of inevitable evaporation of water from the lungs and skin. These unavoidable water losses are known as insensible losses. Human skin is a relatively poor vapor barrier, particularly in hot weather. Water leaves the skin primarily by diffusion at a rate of about 400 milliliters per day in comfortable conditions; at high temperature and low humidity, this water loss may exceed 1000 milliliters per day. Even the skin of a corpse loses water this way; in the arid desert, a human corpse becomes a desiccated mummy within a few days.

The insensible water loss from the respiratory tract normally amounts to 200 milliliters per day, but during moderate activity in a dry climate, this increases to about 450 milliliters per day. The highest rates of respiratory water loss occur during heavy exercise in cold air at high altitude. Cold mountain air contains little moisture, hence the amount of water vapor needed to humidify the inspired air is large. Because exhaled air has been warmed to body temperature, it contains large amounts of water vapor that rains out of the breath. The quantity of water lost by evaporation from the lungs depends directly on the temperature and volume of cold air breathed and may exceed 1.5 liters per day.

Humans are essentially tropical mammals who have evolved the capacity to keep cool by evaporation. Indeed, the capacity to cool by evaporation suggests that humans originated in a hot climate where water was plentiful. The relatively large surface area to mass ratio of the human body and the presence of large numbers of sweat glands make evaporation an efficient mechanism for keeping cool.

I will discuss this evaporative mechanism of cooling in more detail later, but a few comments here are necessary. During sweating water is actively secreted by eccrine sweat glands onto the surface of the skin, where it evaporates. Evaporation consumes heat (latent heat of vaporization), which cools the skin. The water lost by sweating differs from insensible loss in an important way. Sweat is not pure water but contains salt, although at a lower concentration than does blood plasma.

The amount of water lost by sweating depends on environmental temperature and level of activity. In resting, lightly clothed people, sweating begins at an ambient temperature of about 25°C (72°F). During exercise in the desert, sustained water losses from sweating have been measured at 0.5 liters per hour. Peak sweat losses of 1.2 liters in a single hour have been documented in men marching in the desert sun, and during heavy exercise in climate chambers, short-term quantities of 1.5 to 2.5 liters per hour have been reported.

The third problem of lack of water, or more specifically of dehydration relative to starvation, is the absence of a large store of freshwater in the body. It has been noted that a substantial amount of energy can be stored in the form of body fat, which provides six times more energy than does glycogen. In addition to fat, carbon is also obtained from the skeletal muscles, which constitute 40% of normal body weight. The average person thus has many weeks of nutritional reserve, but the body, unlike the camel, for instance, has no significant storage sites for water. The camel's hump is primarily fat, which provides some metabolic water, but the alimentary tract and circulation of the camel hold more than 50 liters of extra water, an amount sufficient for up to10 days of survival under desert conditions. By comparison, a person who drinks his or her fill before setting out across the desert can hold about 2 or 3 liters in the stomach and intestines, an amount that is exhausted in a few hours.

## The Body's Minimum Daily Water Requirements

This background sets the stage for determining the minimum amount of water that must be consumed on an average day to offset obligatory water losses and avoid dehydration. If the obligatory losses are reduced to an absolute minimum and added up, the amounts are 600 milliliters of urine, 400 milliliters of insensible skin loss, and 200 milliliters of respiratory water loss, a total of 1.2 liters. Because maximum urine osmolarity is 1200 milliosmoles per liter (mOsm/L), if diet is adjusted to provide the minimum solute excretion per day (about 600 mOsm), minimum urine output may fall, in theory, to 500 milliliters per day and maintain solute balance. Hence, the absolute minimum water intake amounts to just more than 1 liter per day. This value is not particularly realistic, because it means one cannot sweat a drop, exercise a whit, or have a loose bowel movement without risking dehydration.

The amount of free water in the adult human body varies somewhat from individual to individual, but it averages approximately 60% by weight (mass). Thus, a 70-kilogram man contains roughly 42 liters of free water. This is the total body water, or TBW. The TBW is distributed into two fluid compartments, or body spaces, according to the amount of solute (salt) in each compartment. These compartments are the intracellular fluid (ICF) and the extracellular fluid (ECF) spaces. The ICF contains approximately 55% of the TBW and the ECF about 45%. The ECF space is subdivided further into the interstitial space and the vascular space, or blood plasma.

The volume of plasma is a critical factor in the defense against dehydration. Plasma is the actual fluid component of the circulation. The circulating blood volume amounts to roughly 7% of body mass, and after subtracting the space occupied by red blood cells, the amount of plasma is slightly less than 3 liters. Hence, for a 70-kilogram man with 42 liters of TBW, only one-fourteenth is plasma. Maintaining this plasma volume is critical to stable circulation, and the loss of only 1 liter of plasma will incapacitate an individual because the cardiac output cannot be maintained. Plasma volume must be preserved in hot environments in order to survive for any length of time.

Approximately one-third of TBW is contained in the interstitial space of the body. This space includes the fluid in the lymphatic system, the fluid between cells, secretory fluids (transcellular fluid), and water trapped in cartilage, bone, tendon, and ligament. Most of the fluid in the interstitial space acts as an available reservoir; it is accessible to the other compartments of the body except for the small amount in cartilage and bone.

The distribution of water in the body is determined by the osmotic strength in each compartment. Osmotic strength is related directly to the salt concentration that is "trapped" inside the space. This salt consists of charged molecules, or ions; neither positive (cations) nor negative (anions) ions cross cell membranes without energy input and/or a channel because biological membranes are freely permeable only to water. Such membranes are said to be semipermeable.

The two most important cations in the body are sodium ($Na^+$) and potassium ($K^+$). The positive charges on these ions are accompanied by an equal number of negative charges in the form of anions, such as chloride ($Cl^-$). These ion pairings maintain charge neutrality (electroneutrality) in the body compartments. They also determine the distribution of water in the extracellular and intracellular spaces. $Na^+$ is the primary cation in the extracellular space, while $K^+$ is found primarily in the intracellular space. The concentration of $K^+$ inside the cells of the body is roughly the same as the concentration of $Na^+$ outside the cells. Water moves passively with the movement of these and other ions in the body according to the law of osmosis, by which water is dragged between adjacent compartments until the salt concentration is the same in both.

When the concentrations of all the salts in one body compartment are added up, an estimate is obtained of the osmolar strength, or osmolarity, of the compartment. Sodium and accompanying anions account for the bulk of the active osmotic solute in the ECF. Thus, the volume of the ECF space is determined almost entirely by $Na^+$ content. For instance, in human plasma, in which the concentration of $Na^+$ is approximately 135 milliequivalents per liter, the effect of $Na^+$ accounts for an osmolarity of 270 milliosmoles per liter (mOsm/L); this is 96% of the normal value of 280 milliosmoles per liter. The osmolarity of plasma is tightly regulated by the sensation of thirst. The main stimuli to drink are high osmolarity and low blood volume, and thirst is activated when plasma osmolarity rises to 295 milliosmoles per liter.

Thirst is a critical defense mechanism, but a simple calculation shows it is not activated until after significant dehydration has already occurred. The TBW at which thirst occurs is given by normal TBW × normal osmolarity ÷ by the osmolarity at the thirst threshold. Thus, TBW at thirst is 42 liters × $\frac{280}{295}$ milliosmoles per liter, or 39.9 liters. Therefore, 2.1 liters of body water has been lost before the thirst response is activated. Some people, therefore, live in a chronically dehydrated state just below the thirst threshold. In these individuals the daily flow of urine may be reduced as much as tenfold in order to maintain fluid balance.

## The Mechanism of Dehydration and the Body's Responses

During dehydration the body loses sodium and water, and the ECF volume decreases. Because dehydration is due to insensible losses and sweating, the body always loses water in excess of sodium, and the osmolarity of the ECF increases. This increase in osmolarity is directly proportional to the loss of water from the body. A higher osmolarity causes water from the ICF to move by osmosis into the ECF to maintain the plasma volume. Sweat cannot be replaced by water alone because drinking causes the osmolarity of the ECF to fall by diluting the remaining sodium and other electrolytes. Water intake must be accompanied by an appropriate amount of sodium. In order to regulate this system, the change in both volume and composition of the ECF must be sensed by receptors sensitive to variations in volume and osmolarity.

The body responds to dehydration in two primary ways. First, the remaining body water (TBW) is redistributed between the fluid compartments to ensure that the loss is shared by both the ECF and ICF. If excessive water and salt are lost to the environment, cellular and extracellular dehydration result. Second, the body conserves water primarily by decreasing urine output in direct proportion to the amount of water (TBW) lost from the body. Thus, salt and water loss is compensated for in the kidneys, and water conservation is a major function of all mammalian kidneys.

The kidneys conserve water by excreting urine that is more concentrated than plasma. The concentrating power of the human kidney is approximately 1200 milliosmoles per liter, or roughly four times the osmolarity of plasma. The kidneys concentrate urine by decreasing the amounts of water and sodium excreted. Concentration of urine occurs through the action of water and sodium-retention hormones and the activity of renal nerves under the control of the sympathetic nervous system. Under these influences, the kidneys maintain sodium balance in the ECF compartment. Because ECF volume depends directly on the amount of sodium in the body, the kidneys are the key to a successful defense against dehydration.

By excreting concentrated urine, water is conserved during dehydration, which preserves the remaining ECF volume, particularly the circulating plasma volume. This explains the most obvious manifestation of dehydration: a decrease in urine output in proportion to the loss of body water (TBW). The fall in urine output cannot be attributed simply to a fall in plasma volume; it reflects the net effect of multiple regulatory processes.

Excretion of urine by the kidneys represents primarily the net effect of two large and opposing forces: filtration of the plasma, which normally amounts to about 125 milliliters per minute, and reabsorption of all but 1 milliliter per minute of this filtrate by renal transport processes. Blood flow to the kidneys accounts for about 20% of resting cardiac output even though the kidneys represent only about 1% of body weight. This favorable distribution of blood flow is required for plasma filtration (also called ultrafiltration).

During dehydration blood flow to the kidneys is decreased, which decreases ultrafiltration. The ultrafiltration rate is determined primarily by the amount of blood flowing through the kidneys' thousands of filtering apparatuses, or nephrons. Each nephron contains microscopic bundles of capillaries, or glomeruli, which actively filter the plasma, and slightly less than 1% of this filtrate becomes the urine. The glomerular filtration rate (or GFR) of 125 milliliters per minute may decrease more than 50% from the combined effects of dehydration and heat stress.

During dehydration the salt concentration in the plasma (plasma osmolarity) increases in proportion to the amount of fluid lost. Plasma salt concentration is detected in specialized receptor cells located primarily in the brain, known as osmoreceptors. These osmoreceptors respond to changes in the osmotic pressure of the ECF, in which they are bathed, by shrinking, or contracting. Shrinking stimulates the release of arginine vasopressin (AVP), also known as antidiuretic hormone (ADH), from an area of the brain involving and just above the posterior pituitary gland, the neurohypophysis (see Fig. 3.1). Additional AVP release is stimulated by osmoreceptors in the portal system of the liver and by stimulation of the brain from cardiovascular receptors that sense changes in volume and pressure in the circulation.

The amount of AVP released in the blood increases with the increase in plasma osmolarity and the loss of plasma volume. The osmotic threshold for AVP release is about 285 milliosmoles per liter, and plasma osmolarity is regulated to within 2% of the desired value (280 to 290 milliosmoles per liter). Lack of AVP secretion or action causes diabetes insipidus, a disease characterized by copious production of dilute urine. AVP is released in greater amounts when body temperature is elevated and after heat acclimatization. At elevated body temperature the amount of AVP released is twice that predicted from the increase in plasma osmolarity alone. Thus, AVP provides a critical link between water conservation and heat tolerance.

AVP exerts its effects primarily on the kidneys, which contain receptors in the tubule cells that bind the hormone. The kidneys are so sensitive to this hormone that renal function is affected by AVP infusions too small to be measured after it is diluted in the bloodstream. AVP binds to tubule cells in the kidneys, causing the tubules to reabsorb much greater amounts of water from glomerular filtrate. This water absorption occurs in a fascinating way: AVP binds to a receptor called

V2 on the membrane of cells in the collecting ducts of the nephrons, activating an enzyme system (adenylate cyclase) that directs the nucleus to insert water channels into the cell membranes. Water channels are encoded for by a special gene, *aquaporin-2*, which controls synthesis of the channel protein aquaporin-2. Water channels facilitate reabsorption of water along osmotic gradients from collecting ducts to the interstitial spaces of the central kidney, or renal medulla (Nielsen et al., 2002). This is a key factor in the kidneys' ability to concentrate urine and conserve water during heat stress and dehydration.

The maximal concentrating ability of the kidney during dehydration is not entirely due to the action of AVP because even large amounts of the hormone do not increase urinary salt concentration to the maximal 1200 milliosmoles per liter during profound dehydration. The increase in plasma AVP with dehydration is greater in summer than in winter because of heat acclimatization effects on body fluid balance. During dehydration heat-acclimatized people lose more water and less electrolytes in their sweat, and plasma osmolarity increases more than in unacclimatized subjects. The difference is accounted for by salt-conserving mineralocorticoid hormones released from the adrenal glands after heat acclimatization.

Water conservation during dehydration is assisted by the release of other adrenal hormones, such as aldosterone (controlled by the renin–angiotensin system that regulates blood pressure), and the stress hormone norepinephrine. Dehydration also inhibits release of water anticonservation peptides from the chambers of the heart. When the cardiac atria are physically stretched, several hormones are released into the circulation. One of the most important of these is atrial natriuretic peptide (ANP). ANP and related peptides cause the kidneys to excrete sodium and water. When plasma volume decreases with dehydration, the pressure difference decreases between the inside and the outside of the heart. This decreases cardiac distension and inhibits the release of atrial peptides, thereby decreasing sodium and water excretion by the kidneys.

## Dehydration and Heat Tolerance

This physiological background is the basis of a fascinating story of human adaptation that was unraveled in the early part of the twentieth century. In 1922 K. N. Moss reported a syndrome in deep-earth miners characterized by extreme fatigue, muscle cramps, and disorientation. In some cases the miners actually had epileptic convulsions. This syndrome occurred during periods of exceptionally hard work in hot conditions in coal miners in England.

Within a year Dr. J. S. Haldane had picked up the trail, realizing that experienced miners consumed very little water during their shifts, preferring when thirsty to take a small amount to rinse out their mouths without swallowing it. These men believed too much water in hot conditions was bad for them, and Haldane agreed, referring to the condition as "water poisoning." He attributed it, however, to temporary paralysis of water excretion by the kidney because of the hot conditions. This reluctance to drink was later called "voluntary dehydration" by E. F. Adolf, who recognized that people who work in the heat drink voluntarily only about two-thirds of the water they lose (Adolf, 1947).

Haldane's original analysis of water poisoning was correct as far as it went, but to really understand the problem requires understanding the working conditions of the underground miner. Deep mines are essentially tropical environments, and, for the most part, the deeper the mine the higher the temperature. The deepest mines in the world, the gold mines of South Africa, are more than two miles deep, where temperatures exceed 49°C (120°F) and atmospheric pressure can approach 1.5 times normal. In some mines the rocks are sprayed constantly with water to minimize the aerosolized dust particles that cause deadly lung diseases known as dust pneumoconiosis. Thus, the dense air in deep mines often reaches 100% humidity.

Underground mining is hard and dangerous work, and physical exertion at such high temperatures raises body temperature. Miners sweat profusely, but because of the humidity evaporation of sweat is reduced and cooling by evaporation is less effective than normal. The risks of heat exhaustion and heatstroke are high. Today's miners are often selected on the basis of their ability to acclimatize to work in such extremely hot environments. However, 100 years ago the advantages conferred by heat acclimatization were not appreciated fully.

During physical work in the heat, sweat is composed of both water and salt. Under such conditions the sweat of the nonacclimated worker is nearly isotonic—its salt composition is similar to that of plasma and other body fluids. As pointed out earlier, if only water is replaced, the salt in the body compartments becomes diluted quickly. Over-drinking can cause the plasma sodium to fall to dangerous levels, during which the body suffers adverse effects, such as muscle cramps and convulsions. These are the effects of low sodium, or hyponatremia. Thus, the workers must replace both water and salt at regular intervals, regardless of thirst.

These principles of avoiding the adverse effects of dehydration have been applied to military training and sports medicine; soldiers and athletes exercising in warm weather are trained to drink even if they are not thirsty. Fortunately, the scientifically untenable military tradition of water discipline has been abandoned. However, over-drinking with no salt intake can lead to the undesirable and dangerous consequences of hyponatremia. In addition, plasma volume falls because salt has been lost. Thus, the athlete who collapses on the playing field often presents the physician with a diagnostic dilemma. Depending on what and how much the athlete has been drinking, the problem may be dehydration, hyponatremia, or both. This principle is illustrated in Figure 5.1. If the amount of water intake is unknown, plasma sodium and osmolarity must be measured to determine the appropriate treatment. This concern instigated the development of balanced sports drinks, most of which contain enough sodium to avoid electrolyte dilution.

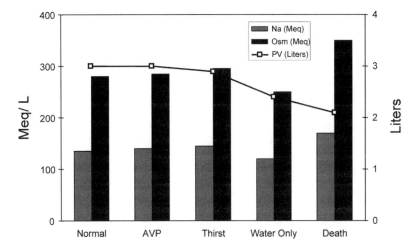

**Figure 5.1.** Progressive changes in plasma volume and composition in response to dehydration. The initial defense against dehydration is secretion of arginine vasopressin (AVP), or anti-diuretic hormone, which allows the kidneys to conserve water and maintain normal plasma osmolarity. Thirst requires a further increase in plasma osmolarity. If only water is replenished, plasma volume (PV), sodium (Na), and osmolarity (Osm) will fall dangerously. Without water repletion plasma volume falls, and the plasma sodium and osmolarity rise to extreme levels.

## Survival Time without Drinking Water

The main purpose of the discussion so far has been to provide enough information to calculate the survival time when no water is available to drink. The problem requires answers to two questions: What is the maximum amount of dehydration that can be tolerated, and what is it about dehydration that kills people? Just as hard and fast answers could not be given to the questions about normal amounts of salt and water in different body compartments and how much water intake is enough, there will also be uncertainty in the answers to these questions about dehydration.

How much body water can be lost by dehydration? It has been estimated that a loss of only 12% of body mass from dehydration can produce clinical shock. For a 70-kilogram person, this amounts to as little as 8.4 liters, or 20% of the TBW. This estimate is based on the highest salt concentration (plasma osmolarity) compatible with life, which is about 350 milliosmoles per liter. In a person with a normal TBW of 42 liters, neglecting sodium and assuming pure water loss, the critical TBW can be determined from normal TBW × the osmolarity ratio, or $\frac{280}{350}$ milliosmoles per liter. This yields a critical TBW of 33.6 liters, or a loss of 8.4 liters. This water loss is apportioned across all fluid compartments according to their size. For plasma the loss would be 12% of 8.4 liters, or approximately 1 liter. This volume is sufficient to seriously compromise the function of circulation.

It should be clear that pure water loss in a hot climate is never possible; sweat is always salty no matter how well acclimatized the individual. Some of the sodium in sweat (about one-quarter) is lost from the plasma, which means the decrease in plasma volume will be larger than if only water were lost because the osmolarity of the plasma will not rise as rapidly. In other words, when sodium is lost from plasma, not as much water can leave the cells by osmosis to maintain the plasma volume. The maximum tolerable decrease in plasma volume is reached sooner the more sodium is lost. Thus, in a 70-kilogram person 8.4 liters is a reasonable upper limit for the maximum tolerable water loss from dehydration. Again, larger individuals with larger plasma volumes can sustain greater absolute water losses.

If a 70-kilogram person can sustain an 8.4-liter water loss, then how long can he or she survive without drinking? Because the minimum daily water loss is fixed at around 1.2 liters, an approximate survival time without water of $\frac{8.4}{1.2}$ liters per day, or seven days, can be estimated This is 168 hours, or 68 hours longer than the often-quoted 100-hour rule of thumb. There are also a few well-documented anecdotes of humans surviving eight to ten days without drinking water, but sporadic reports of survival for up to 14 days without water are highly questionable. For a 70-kilogram person to lose the minimum of 1 liter of water per day for fourteen days yields 14 liters, or 33% of TBW, which is incompatible with survival.

Actual survival time without drinking water depends directly on how much one sweats, which is determined primarily by temperature and humidity and by the amount of heat generated by work, such as walking. Indeed, an individual's behavior in terms of the amount of activity and the conditions under which it is performed will often govern survival time for a given set of conditions. This situation makes it very difficult to predict an accurate survival time without water for a particular individual even if the temperature, humidity, and shade conditions are known.

Despite the lack of precise estimates of survival time without water, a few general rules can be derived on the basis of the physiology of cooling. For a 100-hour period a minimum total urine output of about 2.2 liters must be accounted for, which leaves 6.2 of 8.4 liters to lose as sweat and insensible losses. Ignoring insensible losses, this allows a maximum of 62 milliliters (2 ounces) of sweat per hour, which means minimal activity in warm weather causes an individual to reach the limit. On a daytime desert march, however, it is possible to produce nearly 25 times this amount of sweat (1.5 liters per hour). Furthermore, dehydration, through it effects on decreasing the plasma volume, tends to increase body temperature during exercise, which maintains the water loss from sweating.

In practical terms, at a daytime temperature of roughly 75°F with a minimum of physical activity, expected survival time without drinking water would average 100 hours. As ambient temperature increases, survival time will be reduced in proportion to temperature. At 90°F survival time with limited activity easily can be decreased by a factor of two. If stranded in the desert with little or no water, it is prudent to rest in the shade in daylight and travel at night. It has been estimated that sustained marching in the Sonora desert of northern Mexico at night requires just under a gallon (3.8 liters) of water for every twenty miles, while traveling during the daytime requires more than twice as much.

To date no practical, efficient method of recycling body water has been devised for use under extreme desert conditions. Water recycling is an attractive concept, and simple methods have been developed for collecting urine and sweat and distilling the water using solar energy. The science fiction buff will immediately think of the still-suits of the indigenous desert Fremen of Arachus in Frank Herbert's imaginative novel *Dune*. Any such clothing, however, must either allow the water to evaporate from the skin before it is captured or recirculate the water through a heat exchanger. Otherwise, adequate body cooling will be impossible.

In summary, under extremely hot desert conditions of at least 49°C (120°F), sustained sweat losses of 1.2 to 1.5 liters per hour have been reported during forced marching. Such sustained high sweat rates can reduce estimated survival time without drinking water to as little as seven hours, or approximately the time it takes to walk twenty miles. Because no portable technology is yet available to recycle body water, the desert sojourner has but one survival option: carry a sufficient supply of water to complete the journey on foot under the prevailing weather conditions.

# Water That Makes Men Mad

When Samuel Taylor Coleridge wrote the words, "Water, water everywhere, but not a drop to drink" in *The Rime of the Ancient Mariner* in 1798, the dangers of drinking seawater had been known for thousands of years. Seawater does indeed make men mad. Historical evidence indicates the ancient Egyptians knew seawater was not potable, but the earliest realization that it was unsafe to drink has been lost to antiquity. In pre-Columbian times the greatest fear of venturing too far from land on the ocean was not falling off the surface of the Earth but lack of fresh drinking water. From a human perspective the oceans, which cover 70% of Earth's surface, are still the most extensive and unique desert wildernesses on the planet. Saltwater constitutes 97% of Earth's water, and of the 3% that is fresh, two-thirds is frozen in glaciers and polar ice. Thus, a mere 1% of all the water on the planet (in lakes and rivers, groundwater, and the atmosphere) is fresh and can be used by terrestrial plants and animals.

For a mariner adrift at sea, the most common causes of death are hypothermia and dehydration, and the castaway's fate depends to a great extent on water temperature. Hypothermia is the survival-limiting problem more than 30° N or S of the equator; this safe temperature zone shrinks to 20° N or S in winter. Dehydration is the major problem on the oceans between the tropics of Cancer and Capricorn and in lifeboats and rafts is an important cause of death at all temperate latitudes.

## The Composition of Seawater

A thousand-year-old debate continues to this day about how much seawater, if any, can be ingested safely by human beings. The average salt content of ocean water is 35 parts per thousand (3.5% salinity) and consists of approximately one-third sodium chloride. The other two-thirds of sea salt are chloride, sulfate, and bromide salts of magnesium, potassium, and calcium (Table 6.1). These seven ions make up 95% of the chemicals in seawater, although more than seventy minerals are present in trace amounts. The osmolarity of "standard seawater" is 1035 milliosmoles per liter, or nearly four times that of human blood plasma and close to the maximum solute concentration of urine. Thus, drinking seawater cannot increase body water because its salt content is too high, and it worsens cellular dehydration.

## Ingestion of Seawater

How much seawater can one ingest before it becomes lethal? The practical answer to this question is "None," but the scientifically correct answer is "It depends." It depends on one's state of hydration and how long it will be until fresh water becomes available. In Chapter 5 we saw that water requirements depend primarily on three factors: body size, work, and environment, that is, temperature and humidity. Other variables play smaller roles, such as differences in individual physiology and diet. Although the physiological responses to dehydration are pronounced, evidence is lacking that human beings adapt to water deprivation. Individual, ethnic, and racial differences in tolerance to dehydration are negligible for people of the same size. Acclimation to heat, however, is beneficial because dilute sweat is produced and salt is lost more slowly from ECF, including the plasma.

It seems that human adaptation to water availability is already close to optimal because our daily water requirements are so low. Only 1.25 quarts (1.2 liters) per

**Table 6.1.** Chemical Composition of Seawater

| CHEMICAL ELEMENT | PARTS PER MILLION (PPM) | (%) | MILLIOSMOLES PER LITER |
|---|---|---|---|
| Chlorine | 19,400 | 55.4 | 547 |
| Sodium | 10,800 | 30.8 | 469 |
| Magnesium | 1,290 | 3.7 | 53 |
| Sulfur | 904 | 2.6 | 28 |
| Calcium | 411 | 1.2 | 10 |
| Potassium | 392 | 1.1 | 10 |
| Bromine | 69 | 0.2 | <1 |

day are needed to survive in a temperate climate, which amounts to less than 2% of average body weight. On the other hand, in an open raft at sea in the tropics without a canopy, four to five times this much water may be needed to assure long-term survival. Thus, the issue of drinking seawater usually arises for people adrift in tropical seas without an adequate supply of freshwater. This issue will be dealt with in the context of general guidelines for survival at sea.

## Survival at Sea

The maritime activities of the hundreds of nations with seafaring interests have led to international policies for the practice of survival at sea. Historically, the most important treaty on maritime safety has been the convention for safety of life at sea (SOLAS). The first SOLAS convention in London, attended by thirteen countries, was adopted for commercial shipping in 1914 in response to the *Titanic* disaster. Subsequent conventions in 1929, 1948, and 1960 were attended by larger numbers of participants. The 1960 convention, which came into force on May 26, 1965, modernized ship regulations to keep pace with developments in shipping technology and safety based on knowledge and experience. The procedure for updating it was unwieldy, however, and the 1974 convention, attended by seventy-one countries, allowed for rapid amendment by tacit acceptance. Thus, new amendments entered into force on a prearranged date unless prior objections were received from one-third of the member governments or from governments whose combined merchant fleets represented half the world's gross shipping tonnage.

The 1974 convention has been updated many times, and the document today is referred to as "SOLAS 1974 as Amended." It contains regulations for readiness and safety of ships at sea and procedures for abandonment, survival, detection, and retrieval of survivors of sinking. Chapter 3 of SOLAS, Life-Saving Appliances and Arrangements, includes more than fifty regulations and requirements for lifeboats, life rafts, and personal flotation devices (PFD).

Over the past 25 years amendments have been added for enclosed self-righting lifeboats with engines to limit exposure to the harshness of the open sea. Cargo ships must carry enough enclosed lifeboats and rafts on each side to accommodate everyone on board. On passenger ships, partially or totally enclosed lifeboats along each side are required to accommodate at least half of all persons aboard. Ferries and passenger ships on short voyages can substitute life rafts for some lifeboats. Life-saving appliances must be well maintained and able to be mobilized promptly in an emergency. Survival craft must be capable of being launched with a ship listing 20° in either direction, and cargo ships of 20,000 gross tons or more must be able to launch them while the ship is making headway at up to 5 knots. This requirement acknowledges that modern ships of great size and displacement are difficult to stop.

The advantages of lifeboats and life rafts at sea are so decidedly great as to require no comment. For one who must abandon ship, the most immediate danger to survival is not drowning; professional mariners now know the importance of flotation even if they are strong swimmers. In cold water, however, life vests provide minimal protection against hypothermia. For this threat, SOLAS regulations require immersion suits to reduce body heat loss and thermal protective aids (waterproof bags or suits of low thermal conductivity) for lifeboats and rafts on ships that operate in cold water.

SOLAS life rafts fall into three classes according to the waters in which vessels operate. These rafts (ocean, offshore, and coastal) reflect differences in size, construction, operating features, and requirements for stores and equipment. SOLAS equipment packs are certified separately from rafts and include essential supplies, such as a first aid kits, sea anchors, paddles, bailers, tools, and flares. The SOLAS "A" pack for ocean service requires water, food, thermal protection gear, and a fishing kit, not required in the "B" pack. Interestingly, the "A" pack requires only 1.5 liters of water and 2400 kilocalories of food per person, essentially a twenty-four-hour supply of each. The implication is that even under the most adverse and remote ocean conditions, modern search aircraft and surface vessels and the availability of satellite communications and global positioning systems (GPS) make it possible to accomplish most rescues within a day or two. Modern emergency position-indicating radio beacons (EPIRB) and search-and-rescue transponders (SART) make locating sailors adrift far easier today than in the past, provided the equipment is readily available and in working order.

## Lessons from the USS Indianapolis

In ocean survival plans, the issue of how much drinking water to provide is a vital concern. A SOLAS six-person ocean raft carries only 9 liters of fresh water (2.34 gallons), which will not last long in a tropical climate. As noted earlier, a resting person exposed to tropical sun may require five quarts (4.8 liters) of water per day to avoid dehydration. Failure to equip a raft or lifeboat with drinking water implies that people will be rescued promptly and assumes would-be rescuers will know someone needs to be rescued. If a ship sinks before radioing an SOS or if the batteries in the EPIRB are dead, rescue at sea may be long delayed. In such circumstances failure to store fresh water on rafts is a failure of preparation that may be lethal because no one knows the ship sank.

This exact set of circumstances was faced by the sailors on the *USS Indianapolis*, which was torpedoed and sank in the Pacific four days after unloading the Hiroshima atomic bomb at Tinian Island on July 26, 1945. The ship sank at night in twelve minutes, and no SOS was dispatched. The immediate survivors drifted in the Philippine Sea for four days because the U.S. Navy did not realize the ship was overdue.

58    THE BIOLOGY OF HUMAN SURVIVAL

After departing Tinian, the *Indianapolis* reported to the commander-in-chief, Pacific Headquarters at Guam, for orders. The cruiser was ordered to join the battleship *USS Idaho* at Leyte Gulf in the Philippines to prepare for the invasion of Japan. The *Indianapolis*, unescorted, departed Guam on a course of 262° making about 17 knots. Shortly after midnight on July 30, 1945, the ship was rocked by two explosions from torpedoes fired by a Japanese submarine. Of the 1199 men on board, some 300 went down with the ship. Roughly 900 men were left adrift in the shark-infested waters of the Philippine Sea with a handful of life rafts, most lacking food or water. The *Indianapolis* was not missed, and by the time the survivors were inadvertently spotted from the air more than four days later, only 321 of the crew, spread across more than 100 miles of ocean, were still alive. Ultimately, 317 men survived the terrible ordeal. Apart from some taken by sharks, many who died drank seawater, became delirious, and drifted or swam off, never to be seen again.

It is impossible to know just how many of the men of the *Indianapolis* died strictly of dehydration, because some died of burns or other injuries as well as shark attacks. The first sighting of the survivors was made at 11°54' N and 113°47' E on the morning of August 4, 1945. The surface water temperature at that time and location can be estimated to have been about 30°C (86°F), which means hypothermia, except for the badly burned, would not have threatened the lives of lightly clad men wearing lifejackets in the water.

The accounts of the survivors and rescuers indicate that approximately one-third of those adrift lived for 107 to 120 hours without freshwater and that none of the survivors drank much seawater. The medical officer, who survived the disaster, reported that those who did drink seawater developed diarrhea, became delirious and combative, and then drowned or drifted or swam away. Thus, each survivor was able to adapt behaviorally in the face of one of the most powerful physiological stimuli known, severe thirst, but as the analysis in chapter 5 suggests, an equal number who resisted drinking seawater probably simply ran out of time and died of dehydration before they could be rescued.

The survival problem of the *Indianapolis* disaster is a physiological one. Immersion causes diuresis, which hastens dehydration, and drinking hyperosmolar seawater shortens, not prolongs, survival time. Physiologists know that hypertonic solutions administered into the bloodstream are dangerous because they raise the osmolarity of plasma. The initial increase in osmolarity from such an infusion is calculated as follows. Baseline plasma osmolarity is multiplied by plasma volume and added to the product of infusion volume and infusion osmolarity. The sum is divided by the sum of plasma volume and infusion volume. For example, at a plasma osmolarity of 280 and a plasma volume of 3 liters, rapid infusion of 0.33 liters of a 1000-milliosmoles per liter solution into the circulation gives a plasma osmolarity of [(280 × 3 liters) + (1000 × 0.33 liters)] ÷ 3.33 liters, or 352 milliosmoles per liter! This is a lethal increase in osmolarity.

During a hyperosmolar infusion other mechanisms also enter into play. Water moves out of interstitial and intracellular fluids by osmosis, which prevents plasma osmolarity from rising too fast, and the kidneys immediately begin to excrete the load of salt. This means water moves out of the cells into the circulation and then out of the body via the kidneys as salt is excreted. The net effect is to dehydrate the cells, which shrinks the brain and other organs.

Consider next what happens when seawater is taken into the alimentary tract. Concentrated salt is absorbed into the circulation with water from the intestines. Therefore, the initial effects of drinking seawater are similar to those of infusing hypertonic saltwater directly into the circulation. It temporarily increases plasma volume by drawing water from the cells, which may make one feel stronger for a short time. However, the excess salt must eventually be excreted. In addition, any salt that is not absorbed from the intestines also draws water out of the body into the intestinal lumen and produces osmotic diarrhea.

The human body simply does not have a way to absorb water in lieu of salt without filtering the salt through the kidneys. Salt excretion requires that a fixed amount of water move with the salt even when urine is maximally concentrated. Thus, the greater the salt load, the greater the water excretion, and because the absolute amount of water lost is a critical factor for survival, hyperosmolar salt hastens water loss and shortens survival time. As mentioned earlier, the salt content of seawater is similar to the maximum urinary salt content of severely dehydrated subjects. Hence, net gain of body water cannot occur by imbibing unadulterated seawater, and castaways should not drink it.

A fascinating aspect of this problem can be appreciated by posing a comparative question. Can marine mammals that live at sea for days to a lifetime drink seawater? This is part of the broader question of how marine mammals regulate salt and water balance in a high-salt environment. Indeed, it has been known for 100 years that the blood salt content of most marine mammals does not appreciably differ from that of terrestrial mammals. Thus, like all mammals, marine mammals are osmoregulators.

It is important to realize that water conservation is easier for marine mammals than for their terrestrial cousins because, at least while at sea, they do not use evaporation to keep cool. In fact, many marine mammals have no sweat glands. Diving mammals also minimize water loss from the lungs during foraging by prolonged breath holding, and by extracting a larger amount of the available oxygen from the air in each breath. The use of a fewer breaths means that less water is lost by evaporation from the lungs during a particular activity. This form of periodic breathing, called apneusis, can reduce respiratory water loss by a factor of two.

Some marine mammals have been observed to drink seawater occasionally, including seals, sea lions, and dolphins. Exactly how often this occurs or how much seawater is consumed is not known. Ingestion of seawater is probably incidental in many marine species. Only sea otters, like sea birds and marine reptiles, have been observed to drink seawater regularly. Most marine mammals take advantage of other sources of water, such as that produced by their metabolism or by feeding on substances of about the same or lower salt concentration as that found in their bodies. Thus, sea lions on a diet of fish appear to be able to live without drinking any freshwater at all. On the other hand, whales that subsist on osmoconforming plankton or crustaceans, such as krill, are dependent on foods of salt content close to that of seawater and therefore face the same challenges as if they drank seawater. The ability of the cetacean kidney to conserve water, in the few instances in which it has been measured, appears to be slightly better than that of humans but not sufficiently so as to convincingly conclude that whales can rely on seawater to maintain water balance.

In sea otters and some species of sea lions and seals, the kidneys are specially adapted to life in a marine environment, as the animals can produce urine that is more than twice as concentrated as that of humans and contains more sodium chloride than seawater. The kidneys of these animals can therefore effectively extract freshwater from seawater. The exact structural and molecular mechanisms of this adaptation have not yet been determined (see Table 6.2).

## A Practical Approach to Salt and Water Loss at Sea

Because the main problem with lack of potable water at sea is dehydration, many points discussed in the previous chapter are relevant to survival at sea. As noted, pure deficiencies of either water or salt are extremely rare. Sodium deficiency occurs in association with sweating and can rapidly lead to profound depletion of the ECF and plasma volume, thus incapacitating an individual. Seawater is certainly one way to replenish a sodium deficiency, and sips of seawater have been recommended as a folk remedy for seasickness for centuries. Can seawater be diluted with fresh water to provide just the right balance of salt and water for the body? In theory this idea could be used to extend the water supply and lengthen

Table 6.2. Maximum Concentrating Ability of Mammalian Kidneys

| TYPE OF KIDNEY | URINE OSMOLARITY (MILLIOSMOLES PER LITER) |
|---|---|
| Sea otter | 2250 |
| Baleen whale | 1350 |
| Human | 1200 |
| Camel | 2800 |
| Sand rat | 6300 |

survival time, but the problem is a practical one. The castaway must know exactly how much salt is in seawater and must make the dilutions so accurately as not to leave the salt concentration too high and spoil a precious supply of freshwater.

For the sake of argument, assume seawater has an osmolarity of 1100 milliosmoles per liter. In principle, then, seawater must be diluted from 1100 milliosmoles to 294 milliosmoles to avoid making thirst more severe or overconcentrating the plasma. Thus, for every ounce of seawater, 2.7 ounces of freshwater must be added to make it potable. In principle, this would forestall the deficiency in plasma sodium and theoretically extend survival time for a given amount of freshwater by a factor of $\frac{(3.7-2.7)}{2.7} \times 100$, or 37%.

This plan is perhaps too uncertain for people adrift on a life raft. What are the alternatives for castaways to extend water resources and obtain freshwater? There are three traditional rules, and they are very simple. First, avoid seasickness, with medication if necessary, because vomiting leads to water and salt loss and hastens dehydration. Second, no one should drink water until they develop a strong sense of thirst. Thirst means plasma osmolarity has risen to the thirst threshold and the body's production of the water conservation hormone arginine vasopressin (AVP) is high. Thereafter, drinking half a liter a day doubles survival time in temperate weather. This strategy buys time to obtain freshwater. Finally, freshwater must be collected and stored at *every* opportunity.

Two obvious sources of freshwater come to mind in this situation: precipitation and floating ice. Rain should always be collected in as large a quantity as is possible and stored in every available container. In cold seas sea ice can provide drinking water under certain conditions. When saltwater freezes, crystalline irregularities cause enough salt to leach slowly out over several years to make it safe to consume. Such "aged" sea ice, also called multiyear ice, tends to be bluish in color but must be melted and the water tasted or tested to determine if it is safe to drink. Sea ice should not be confused with ice from icebergs, which, although less common than frozen seawater, is glacial water and safe to drink.

The consumption of fish and other sea life has been an area of controversy in the literature on survival at sea. Most fish and other sea creatures contain water with little salt in their flesh. The water can be pressed out and collected or the fish eaten immediately. However, the blood of saltwater fishes contains too much salt to be eaten safely. This is not true of birds and turtles, whose blood is a good source of potable water. Life rafts are not routinely equipped with portable solar stills or hand-operated reverse-osmosis desalinators, but these devices are inexpensive, are readily available, and can be added to emergency equipment packs.

The issue of whether to eat when one lacks drinking water has also been a point of contention. All foods except highly dehydrated rations contain reasonable amounts of water. The problem is that protein-rich foods contain amino acids, the ammonia from which is converted by the liver to urea. Urea is the major solute in the urine, and although it is critical to the ability of the kidneys to concentrate urine, it must be excreted with water. If more urea is synthesized, there is more to excrete, the kidney filters more of it, and more water is lost. In other words, the body has conflicting needs when it comes to conserving water and excreting urea. The general point is that if a food contains less water than is needed to excrete its by-products, then eating it increases the net loss of water.

This interesting problem in metabolism unfortunately has not been studied in sufficient detail to make specific recommendations about how much protein is safe to consume when water is limited. In general, in the absence of drinking water, high-protein foods should not be eaten because they hasten water excretion. On the other hand, sugars and other carbohydrates are converted to carbon dioxide ($CO_2$) and metabolic water, which can amount to 300 milliliters per day. The $CO_2$ is exhaled in the breath, and except for the small amount of respiratory water lost in the extra ventilation, it does not significantly increase the net loss of water. Carbohydrate fuels also prevent the body from consuming its own protein, for example, skeletal muscle, which ensures that urea production and urinary water loss are kept to a minimum. Thus, ocean survival rations should consist primarily of nondehydrated carbohydrates. In the Royal Navy, barley sugar, which is 96% carbohydrate and provides 396 calories per 100 grams, has been a standard ration for life rafts since World War II. The perfect food for such an eventuality has often been considered the date, which contains 15% water, 60% carbohydrate, a trace of sodium, and almost no protein.

# 7

## Tolerance to Heat

The dispersion of heat is driven by the universe's thermodynamic imperative, against which rage the furnaces of animal homeothermy. Body heat offers the scythe of speed to animals of advanced phylogeny. In the current animal kingdom, however, only birds and mammals are able to maintain a body temperature that is substantially warmer than their surroundings. Lower animals are cold-blooded, or poikilothermic, which means internal body temperature fluctuates with ambient temperature. Warm-blooded animals are homeothermic because they maintain a constant internal body temperature (core temperature) that is higher than the average temperature of the environment in which they live. This characteristic, which in cooler climates has a strong survival advantage, results from the ability to conserve the metabolic heat every cell in the body generates during the process of homeostasis.

### Mammalian Homeothermy

Most warm-blooded animals maintain a resting internal body temperature within a very narrow range, usually within 2°C or 3°C of basal temperature. This homeothermy is maintained at a wide range of environmental temperatures both above and below the regulated temperature. Tight regulation of body temperature is common to birds and mammals, including humans, and it requires an advanced thermoregulatory system (Hensel et al., 1973). More sophisticated thermal strategies designed to conserve energy, such as estivation and hibernation, have also evolved in some species of homeothermic animals.

For humans the normal body, or core, temperature is 37°C, with a diurnal variation of plus or minus 0.5°C. A temperature below 35°C defines the onset of hypothermia. At 30°C the human body stops generating heat by shivering and assumes the temperature of the environment; it becomes poikilothermic. Death from cardiac rhythm disturbances (ventricular arrhythmia) can occur at any time. On the other hand, a temperature above 38°C at rest is considered fever unless it has been caused by exercise or heat exposure. During sustained heavy exercise, body temperature may reach 40°C. However, temperatures above 41°C can cause heatstroke, which may be fatal. The relationship between body temperature and fatal heatstroke is illustrated in Figure 7.1. The death rate approaches 100% when the body temperature exceeds 44°C. Thus, the range of body temperatures compatible with human life is astonishingly narrow, approximately 30°C to 41°C. On the absolute temperature scale, on which normal body temperature is 310K and the homeostatic range is 303K to 314K; this variance is only 3.5%!

## Humans as Tropical Primates

Humans, like most other primates, are essentially tropical creatures; people are naturally most comfortable in warmer climates and acclimatize better to heat than

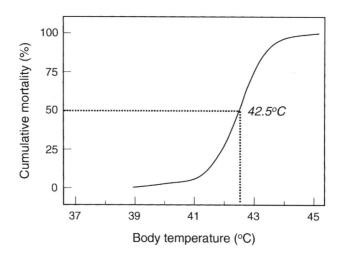

**Figure 7.1.** Human temperature–response curve for cumulative mortality versus body temperature on admission of heatstroke patients to the hospital. Notice that half the victims with body temperatures of 42.5°C or above will die despite treatment. (Data averaged from multiple sources, and curve fit by computer software.)

to cold. Humans rely largely on physiological responses to adapt to heat and on behavior to adapt to cold. People live productively in cooler climates only by changing their behavior. These behavioral adaptations include wearing clothing, building shelters, and maintaining sources of external heat.

Humans tolerate cold quite poorly compared to most other mammals (Feist and White, 1989). For instance, the champion cold weather animal, the Arctic fox, has a pelt with which it can maintain thermal balance without shivering at temperatures as low as –40°C. In contrast, the ambient temperature needed for unclothed humans to maintain thermal neutrality at rest is approximately 25°C (75°F). If this seems high, perform the following simple experiment on yourself. Sleep one night in a room at a temperature of 21°C (70°F) without wearing pajamas or covering up with a sheet or blanket. Unless you are unusually fat, you will awaken shivering within an hour or two.

If people are tropical creatures who fare better in hot than in cold climates, it is natural to wonder how much better humans adapt physiologically to heat than to cold. Healthy people, given access to shade and water, can acclimate to extended heat stress of virtually any climate on the planet. The dangers of heat stress occur primarily in poorly acclimatized people, including children and the elderly, during exercise in the heat and in occupations such as firefighting and deep mining, where heavy protective clothing or closed environments make heat stress dangerous. On the other hand, people who live more than 20° above or below the equator need clothing and other forms of thermal protection to keep warm. This simple observation makes it clear that *Homo sapiens* have lived in equatorial environments for thousands of years, and, as the fossil evidence indicates, originated there.

## Body Heat Balance

All this suggests essential differences between the mechanisms of physiological adaptation to cold and those of adaptation to heat. To understand how humans adapt to cold and hot climates, something must be known about body heat balance and temperature regulation. Fortunately, principles of body heat exchange apply to both extremes of temperature. Once the physical basis of heat exchange is understood, some sense can be made of the physiological responses that lead to adaptation to extremes of temperature.

Heat transfer to or from the human body obeys the second law of thermodynamics. This means that heat transfer from a system of higher temperature to one of a lower temperature occurs at a rate directly proportional to the temperature difference between the two systems. Thus, whenever the temperature of the skin is warmer than ambient temperature, the body dissipates heat; whenever skin temperature is cooler than ambient temperature, heat is transferred from the surroundings to the body. Because the shape and structure of the human body are complex, physiologists have developed detailed models that fill entire textbooks to describe the physics and physiology of heat exchange, temperature regulation, and thermal sensor mechanisms. However, most of what needs to be known about surviving heat stress can be derived from the elementary physics of heat exchange.

Body heat exchange occurs by four physical processes, or modes, conduction, convection, radiation, and evaporation. All four modes of heat transfer are important in heat balance in dry environments, while all but evaporation are important in wet environments. For conceptual and practical reasons, it is important to understand the differences among the four modes of heat transfer.

Heat transfer between two systems in direct physical contact, such as between two bodies in a sleeping bag or when a finger is burned on a hot serving dish, is conduction. Heat transfer in moving fluids or between a still body and a moving fluid surrounding it, such as air or water, is convection. Most of the time, conduction and convection are linked. For example, during immersion of a hand in a cold running stream, heat is transferred by conduction to the water touching the skin, and then it flows away by convection. The volume of water in the stream is huge compared to the volume on the surface of the hand, and convection accounts for most of the heat lost from the hand. This is, of course, also true of immersion of the whole body in flowing water.

Another example of convection is that of wind chill in cold weather. Wind chill accounts for the difference between the rates of skin cooling at a constant temperature in still air compared to moving air. As wind velocity increases more heat leaves the body at any temperature by convection (and evaporation). Not only does wind chill affect one's sense of thermal comfort, the sensation is accurate in terms of body heat balance. The wind chill temperature index (WCTI) actually predicts the risk of cold exposure injury, such as frostbite.*

The transfer of heat between two bodies at a distance is radiant heat exchange, or radiation, also known as blackbody radiation. In other words, the second law requires that heat be transferred between two bodies of different temperatures across a distance provided they are in direct line of sight of one another. Hence, the Sun feels warm because it radiates heat to the body across the near vacuum of space.

The heat required to overcome the change in physical state when liquid is converted to a vapor at constant temperature is evaporative heat exchange, or evaporation. Evaporation is essential to cool the body in hot environments. As sweat evaporates, it takes heat with it, thereby cooling the body. The evaporation of one

---

*In 2001, the National Weather Service adopted a more accurate wind chill formula: WCTI = $35.74 + 0.6215T - 35.75 (V^{0.16}) + 0.4275T (V^{0.16})$ where T = temperature (°F) and V = wind speed (in miles per hour).

gram of water at 30°C requires 2.43 kilojoules (kJ) of energy. This energy is the latent heat of vaporization.

Evaporation is also responsible for most of the cooling of the respiratory tract when breathing cold air. Such respiratory heat losses are significant at high altitude, when the rate and depth of breathing are stimulated as altitude increases and temperature falls. Respiratory heat loss is also a problem in very deep diving, but convection, which increases directly with the density of breathing gas, supplants evaporation as the major component of respiratory heat loss.

Heat exchange between a body and its environment is determined by simply adding the amount of heat from each mode of exchange to the metabolic heat production. This takes the form of a heat balance equation:

$$M \pm C \pm R - E = S$$

where M is metabolic heat production, C is convective plus conductive heat loss (or gain), R is radiant heat loss (or gain), E is evaporative heat loss, and S is net heat storage. The values are most often expressed in units of energy divided by time or power (watts). The positive and negative signs indicate whether heat is lost (negative) or gained (positive). In a living body M is always positive and E is always negative (or zero). M is the energy liberated by the chemical reactions involved in the normal metabolic activities of cells. In a 70-kilogram person at rest, M is roughly equivalent to the power of a 100-watt light bulb and with exercise can increase more than tenfold.

Metabolic heat that reaches the surface of the body must leave by convection, radiation, or evaporation from the skin. In addition, any work done against an external force must be added into the heat balance equation to account for the extra heat generated. This is because muscular work is inefficient, and muscles generate extra heat during contraction; for most activities human efficiency is 5% to 25%. Thus, dissipating the heat generated by muscular work is an important physiological problem, particularly in hot environments.

Heat losses from the skin and respiratory tract contribute to both evaporation and convection in the heat balance equation. In animals that pant, such as birds and dogs, the respiratory component is the primary cooling mechanism. In humans and other primates, however, convective losses from the skin ($C_{sk}$) account for most of the heat lost from the body. At high ambient temperatures, when convective heat loss is no longer possible, evaporation of sweat takes over as the only cooling mechanism. The relative contributions of panting and sweating to body cooling vary enormously from species to species (Figure 7.2). Large tropical animals tend to keep cool by sweating rather than by panting because, ultimately, more heat can be dissipated from the surface of the skin. This mechanism of cooling naturally would have been preferred in tropical animals because it confers a survival advantage in the heat, whereas a heavy

68   THE BIOLOGY OF HUMAN SURVIVAL

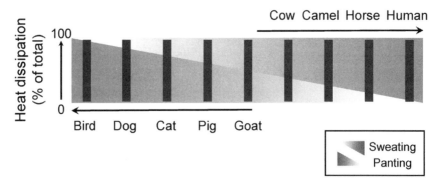

**Figure 7.2.** Reciprocal relationship between cooling by panting or sweating in various warm-blooded animal species. Panting allows cooling by conduction and convection when air temperature is less than body temperature. Sweating cools by evaporation. When air temperature is above body temperature, both strategies keep the body cool by evaporation. (Adapted from Hensel, Herbert, Kurt Brück, and P. Raths. 1973. Homeothermic organisms. In: *Temperature and Life*. Edited by Herbert Precht, J. Christophersen, Herbert Hensel, and Walter Larcher. New York: Springer-Verlag, pp. 503–532.)

coat confers a stronger advantage in northerly and southerly latitudes, where it is much colder.

Convection causes most body cooling during exposure to cold air or cold water and dissipates heat during exercise. The body regulates the amount of heat that reaches the skin by regulating the blood flow to skin and muscles. In thermal physiology, the peripheral region of blood flow regulation is often referred to as the body shell. In the shell skin and muscle blood vessels constrict in cold and dilate in warm conditions to conserve and dissipate heat, respectively. Regulating peripheral blood flow by vasoconstriction and vasodilation is an important part of what allows internal, or core, temperature to remain constant at a wide range of ambient temperatures. A diagram of the shell and core thermal model of the body is shown in Figure 7.3. This simple but useful model was first suggested in 1958 by the German physiologists J. Aschoff and R. Wever.

It should be mentioned that body motion, like fluid motion, alters heat exchange by convection and evaporation. In cold environments body motion actually increases convective heat loss, offsetting to some extent heat produced by exercise or shivering. A good example is the heat generated by swimming in cool water. The principle is identical to cooling by wind chill as described earlier or the cooling effect of a pleasant breeze on a hot day.

To fully appreciate the principles of convection only requires some algebra. Convection from the skin surface of the human body ($C_{sk}$) can be expressed simply as a constant × a temperature difference:

$$C_{sk} = -h_c (T_{sk} - T_a)$$

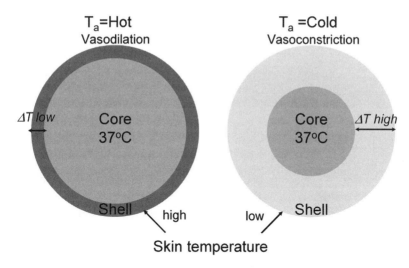

**Figure 7.3.** Simple body core and shell model of human heat management. The size of the body shell decreases in hot environments (left) and increases in cold environments (right). At low temperatures shell thickness is increased by constriction of blood vessels of the skin, subcutaneous tissue, and skeletal muscle. At high temperatures shell thickness is decreased by dilation of vessels of the skin and subcutaneous tissues. If ambient temperature exceeds body temperature, the body gains heat unless sweat can evaporate.

where $h_c$ is a convective heat transfer coefficient, $T_{\overline{sk}}$ = mean skin temperature (°C), and $T_a$ is ambient temperature (°C). The means of deriving a mean skin temperature is hotly debated among thermal physiologists because of the problems of how to weight the skin temperatures of different regions and how finely to divide the regions to compute an average for the entire body. This problem is most difficult in cold environments, where differences in regional skin temperatures are larger than in hot environments, where skin temperatures of all parts of the body tend to be similar. Fortunately for the purposes of this discussion, the concept of mean skin temperature can be used without concern for these details.

The convective heat transfer coefficient ($h_c$) is usually expressed in power per unit of body surface × temperature ($\frac{W}{m^2 \times °C}$). The coefficient is determined by the velocity, density, and heat capacity of the fluid (gas or liquid) around the body. Thus, heat transfer coefficients change with ambient conditions. A negative sign indicates loss of heat when ambient temperature is lower than skin temperature. As will be seen, heat transfer coefficients are important in determining heat exchange in all types of environments, including the near vacuum of deep space, where they approach zero.

In hot environments $C_{sk}$ can actually become positive. In other words, when ambient temperature is warmer than skin temperature, the term in parentheses in the previous equation is negative, $C_{sk}$ becomes positive, and the body gains heat.

If work is performed under such circumstances, metabolic heat production can no longer be dealt with by convection, and body temperature rises. Under these conditions the body has no choice but to rely on evaporation for cooling. This demonstrates why sweating is indispensable for work in hot environments and why it is critical to acclimatization to heat.

## Heat Acclimatization

Heat acclimatization is accompanied by a constellation of adaptive responses that develop over a period of one to three weeks during continuous exposure to hot conditions. These responses are maintained throughout the exposure and disappear over a period of about three weeks when the individual returns to cooler conditions. The essential physiological features of heat acclimatization include five major responses, two of which involve sweating.

The sweat responses of heat acclimatization are the rapid onset of sweating and the production of dilute sweat. The advantage of dilute sweat is that less sodium is lost from the body. For instance, unacclimatized people may secrete sweat with a sodium concentration of 60 milliequivalents per liter while fully acclimatized individuals produce sweat as dilute as 5 milliequivalents per liter. This remarkable conservation of salt primarily involves the action of the salt-conserving adrenal hormone aldosterone on the sweat glands.

The third major response to heat acclimatization is that of plasma volume expansion. Many of the biochemical effects of acute heat stress involve significant elevations in the circulating levels of fluid-conserving and electrolyte-conserving hormones, which subside gradually as acclimatization nears completion. These effects increase sodium and water retention by the kidney, which increases ECF and plasma volume. It is also interesting to note that the concentration of protein in the plasma increases during heat acclimatization. This, too, contributes to the increase of plasma volume. However, the exact mechanism by which it occurs is not known.

The two remaining major responses of heat acclimatization are a small increase in heart rate and an increase in body temperature during exercise. These responses appear to be primarily a consequence of the expanding plasma volume. All five responses facilitate the ability to lose heat from the skin by both evaporation and convection. Furthermore, they allow acclimatized individuals to perform more work in the heat because optimal mechanisms for protection against dehydration are in place. During prolonged exercise in the heat, the decrease in plasma volume caused by dehydration correlates with degradation of athletic performance (Figure 7.4A) and with rise in body temperature (Figure 7.4B). After approximately a week of daily exercise in the heat, most individuals have adapted sufficiently to perform the same level of exercise for the same duration as they

Tolerance to Heat    71

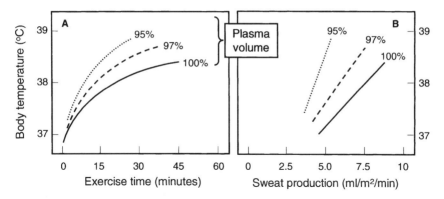

**Figure 7.4.** Effects of dehydration on duration of exercise and body temperature. Dehydration impairs exercise performance primarily by decreasing the maximum exercise time (panel A). Body temperature rises more rapidly in the heat during progressive dehydration because of recruitment of mechanisms for water conservation (less sweating). This causes less effective dissipation of heat from the skin (panel B).

did before heat exposure (Figure 7.5). However, there is considerable individual variability in the time course and extent of heat acclimatization.

## Heat Acclimatization and Physical Fitness

The classic signs of heat acclimatization have sometimes been associated with exercise training in the heat. These signs are a lower heart rate, a higher sweat production, and a lower body core temperature during constant exercise. Exercising muscles readily reach temperatures of 44°C, which is enough to induce heat stress proteins. This raises the question of whether improving physical fitness actually improves tolerance to heat, such as through aerobic or endurance training. During aerobic exercise in temperate climates, many endurance-trained athletes show the signs of heat acclimatization. However, a high level of aerobic fitness is not always associated with improved heat tolerance. Most of the evidence suggests that training sessions must be accompanied by significant elevations of body core temperature and rate of sweating for endurance training to improve heat tolerance. The physical training approach to heat tolerance, however, does work to some extent. If body temperature increases repeatedly by 1° or 2°C during prolonged training sessions, some of the body's heat acclimatization responses will be activated. In addition, training in hot weather tends to decrease the thickness of subcutaneous fat, which improves heat tolerance. Thus, highly fit endurance athletes tend to have excellent heat tolerance, but heat acclimatization alone does little, if anything, to improve exercise fitness.

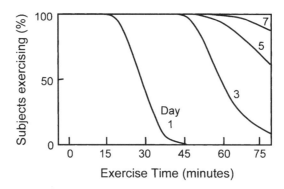

**Figure 7.5.** Improved exercise duration at a constant workload after acclimatization to heat. The effects of heat exposure on exercise are illustrated by gradual acclimatization to a standard exercise regimen in hot conditions. The acclimatization response to exercise occurs in most normal subjects within about one week. Other responses required for full acclimatization may take an additional one to two weeks.

## The Limitations of Human Tolerance to Heat

This discussion raises the issue of what limits human tolerance to heat. It has been emphasized how important salt and water are for life and how important expansion of the plasma volume and changes in sweat composition are for heat acclimatization. These factors are critical in hot climates because humans depend completely on cooling by evaporation whenever the air is hotter than the skin. Under such conditions, unless it is able to sweat, the body will gain heat. Furthermore, the higher the air temperature and the heavier the workload, the faster the body gains heat. Therefore, duration of tolerance to heat declines with increasing amounts of physical work.

The skin temperature that most people find comfortable is approximately 33°C (91.4°F). This is known as the thermoneutral point, and the fact that it is relatively high is a function of our tropical origin. In hot conditions and during exercise, skin temperature rises due to vasodilation of blood vessels, which allows transfer of heat by convection from the body core via the blood to the shell, where it is dissipated to the environment. Internal body temperature is normally 37°C. Hence, when skin temperature reaches 37°C (or whatever internal temperature happens to be), keeping cool depends entirely on sweating. Anyone who has spent a short time in a sauna at a temperature above 37°C has experienced this effect.

As the discussion indicates, heat stress is a thermal load imposed by air movement, temperature, humidity, and body heat generation. Heat stress in deep mines and other hot work environments, including hot weather military operations, has led to the development of techniques to assess the degree of stress and provide guidelines for excessive heat exposure. These techniques attempt to account for factors that impose heat loads on the body in different environments, but because of the complexity of the problem no optimal definition exists for heat stress.

A temperature of 38°C is widely accepted as a safe upper limit for the human body, but in practice it is impractical to monitor body temperatures in large groups of people. Therefore, a variety of heat stress indexes that rely on mathematical combinations of temperature measurements have been used to determine effective temperature, that is, one that predicts the heat strain on the body. It was J. S. Haldane who first proposed in 1905 that the simple ambient (dry bulb) temperature was not a good indicator of physiological heat strain. Although many heat stress indexes can be found in the literature, the simplest and most widely used has been the wet bulb globe temperature, or WBGT. The WBGT index is the weighted sum of three temperature measurements, and it makes Haldane's point rather nicely:

$$WBGT = 0.7T_{nwb} + 0.2T_g + 0.1T_{db}$$

where $T_{nwb}$ is the natural wet bulb, $T_g$ is globe, and $T_{db}$ is dry, or air, temperature. The wet bulb measurement, which requires a wet covering, or wick, around the temperature probe, accounts for evaporation and air movement. Limits for the WBGT index have been established by the National Institute for Occupational Safety and Health (NIOSH) for unacclimatized and acclimatized workers that vary with the average hourly metabolic rate of the work. In addition, limits have been established for which exposure-appropriate protective clothing and equipment is required. These temperature ceilings are in the range of 34°C to 40°C WBGT (Ramsey and Bernard, 2000).

## Heat Illnesses

The human body can store an amount of extra heat that allows its temperature to rise about 5°C before irreversible cellular damage, known as heatstroke, occurs. This means the safe upper limit for internal body temperature is roughly 41°C (106°F). As body temperature increases, this also becomes the maximum tolerable skin temperature and is why immersion in hot water at a temperature above 41°C, which prevents cooling by evaporation, eventually causes death from hyperthermia.

That heat tolerance depends to a large extent on the absolute quantity of heat gained by the body illustrates why it is difficult to predict survival time from ambient temperature alone. The amount of heat needed to raise body temperature above 41°C varies from person to person and depends on the size of the person, the extent of acclimatization, and the individual's behavior in the heat. For example, a child who weighs only 10 kilograms (22 pounds) may die of heatstroke if left in a hot automobile for as little as fifteen minutes. This problem arises because the mass of the child is so small and its ability to keep cool by sweating is more limited than that of an adult. The core temperature of a small child rises even faster than would be predicted from its mass relative to that of an adult because the ratio of body surface area to volume is twice as great in the child. Thus, whenever the temperature in the passenger compartment is greater than 41°C, the child is at risk of heatstroke, and the passenger compartment of an automobile can reach 60°C (140°F) in summert even with the windows partly rolled down.

A graphic example of this process is that of cooking a turkey in an oven. A small turkey cooks faster than a larger one at the same oven temperature because it has a lower mass and a greater surface area to volume ratio (Anderson, 1999). The heat gain is related to the exposed surface area, and the smaller mass allows the internal temperature of the smaller turkey to increase more rapidly. Unfortunately, the same physics applies to a child's body. In the past decade, hundreds of young children in the United States have died of heatstroke after being left unattended in automobiles in hot weather.

An important factor in heat tolerance, as noted above, is the rate of heat gain. The more rapid the heat gain, the higher the incidence of heatstroke. The rate factor makes it difficult to predict survival time from either the absolute amount of heat gain or the body temperature, but this also explains why unprotected people are able to tolerate extremely high temperatures for several minutes. Indeed, an adult can survive in a 150°C (302°F)-oven for ten minutes without a protective garment because the critical gain of heat and loss of water take time.

When ambient temperature is greater than body temperature, the rate of body heat gain depends on only two factors: the rate of heat production (M) and the rate of water evaporation (E). Even respiratory tract cooling becomes purely evaporative, and the heat balance equation reduces to

$$S = M - E$$

This is part of the reason why heatstroke occurs so rapidly if heat loss by evaporation cannot occur, for example, while wearing occlusive garments or after sweating fails. It also illustrates why it is important to dry the skin periodically during exercise in extremely hot weather. Drying helps maintain cooling by evaporation.

The reasons people die in the heat are complicated because heat stress syndromes involve near-simultaneous failures of multiple physiological and biochemical processes. However, a large body of fascinating new information has become available about the molecular events that lead to cellular injury and cell death during hyperthermia (Bouchama and Knochel, 2002). These events have much to teach about adaptation and survival.

When people fail to adapt to heat, manifestations of heat-related illness appear that reflect a continuum of breakdown of physiological and biochemical processes by thermal stress. For practical and traditional reasons, heat illness is usually classified by severity, although the features of various clinical syndromes overlap. The mildest syndrome takes the form of heat cramps, painful muscle spasms that accompany physical work in the heat. Heat exhaustion is marked by headache, lassitude, weakness, and gastrointestinal symptoms such as nausea, vomiting, and diarrhea. These maladies are self-limiting; they improve with rest, fluids, and cooling down. The most severe form of heat illness, heatstroke, is a medical emergency often attended by a triad of symptoms: incoordination, hot dry skin, and a high internal body temperature (hyperpyrexia). Body temperature is usually greater than 41.1°C (106°F), and, more importantly, brain temperature is greater than 40.6°C (105°F). In severe cases the patient may be delirious, comatose, or have epileptic seizures. This triad is present in only about half of patients and occurs more often in classic heatstroke, that is, related strictly to heat stress. It is less common in exertional heatstroke. In either case heatstroke represents a potentially lethal failure of thermoregulation.

Heatstroke is the main reason most people die of heat exposure. In the United States, an average of 250 people die of heatstroke every summer. However, during prolonged heat waves the death toll can be much higher (Smoyer, 1998). For instance, in the summer of 1980, more than 10,000 people in the United States died of heat-related illnesses, and in the heat wave of 1988, the excess number of deaths was 5000 to 10,000 people. If global warming continues the number of heat-related deaths each year can be expected to rise, particularly in densely populated urban areas with many elderly and poor inhabitants. Large cities serve as massive heat islands, particularly at night, when buildings and pavement heated in the day continue to radiate heat. Human studies have shown that urban residence, in addition to advanced age and lower socioeconomic status, are risk factors for death from heat stress. Furthermore, heat can provoke or aggravate many preexisting medical problems, and pre-existing illnesses and infections predispose to heatstroke.

There are many venues where heatstroke may occur, but one setting is more deadly than the rest. This occurs when ambient temperature is greater than body temperature and humidity and radiant (solar) heat load are high. These conditions are considered especially dangerous for heatstroke, sometimes referred to in this setting as sunstroke. Concern always rises when evaporation is the only means of dissipating heat. Profuse sweating hastens dehydration and keeps the skin wet, which impairs evaporative cooling, and lack of acclimatization hastens salt loss and accelerates decline in plasma volume. This situation puts enormous stress on the cardiovascular system. Among the most dangerous of such conditions is encountered inside the notorious rubber exercise suit, which can rapidly induce heatstroke if worn during exercise in even moderate heat.

Once evaporative cooling begins to fail, body temperature rises quickly and circulatory compromise soon appears. The distribution of blood flow to the skin and the high rate of sweating so critical to heat dissipation are overcome by the physical action of heat on thermal regulatory responses. Thus, the first step in homeostatic failure follows from the effects of heating on specific molecular processes in the cell. An example is the effect of temperature on the firing rate of excitable cells in the nervous system that help regulate the circulatory responses to heat stress. Many such cells, especially in the spinal cord, discharge at an increased rate with increasing temperature until they reach 40°C to 41°C, then the firing rate begins to fail. Many effects of heating contribute to this failure, including detrimental changes in the fluidity of cell membranes and the function of ion pumps. Some cellular effects of heat, such as protein denaturation, are opposed by compensatory induction of protective stress proteins as the cell tries to adapt. In any event, these biochemical responses, balanced by the rate of heating, ultimately determine the extent of adaptation and the ability to survive extreme heat stress.

## Death by Heatstroke

The most dangerous and often lethal effects of heatstroke are shock and multiple organ failure. The shock of heatstroke is very similar to another potentially lethal form of shock, septic shock, most often caused by a blood-borne infection from one of the many types of microbes that populate the intestines. These microbes are known as gram-negative bacteria because their cell walls do not take up a dye called Gram stain. In the 1980s the story that endotoxins in the walls of gram-negative bacteria were involved in heatstroke began to be pieced together. The primary culprits are lipid–sugar compounds called lipopolysaccharides (LPS), of which thousands are known. However, normal mechanisms exist to safely sequester these molecules inside the intestinal lumen because LPS causes circulatory shock and organ failure if it enters the circulation at more than minute concentrations.

Most heatstroke victims show rather dramatic increases in the plasma concentration of LPS, which correlates with the development of shock. Indeed, athletes who perform heavy exercise in the heat show increases in plasma LPS concentration that degrade their performance and that appear to be neutralized by antibodies to LPS that their immune systems produce while in training. LPS activates inflammatory cells and causes them to release many cytokines, including tumor necrosis factor (TNF) and the endogenous pyrogen IL-1, that lead to additional elevations in body temperature. The common features of the systemic inflammatory response to heatstroke and septic shock suggest that they cause organ failure by similar mechanisms. The main difference between heatstroke and septic shock is that brain dysfunction (encephalopathy) tends to be more severe in heatstroke.

If LPS contributes to organ failure and death from heatstroke, where does it come from, and how does it enter the bloodstream? The question of where it comes from has been answered: the intestinal lumen. The answer to how it moves from the intestinal lumen into the blood is very interesting. Normally, the intestinal lining forms a tight barrier to the movement of foreign molecules. The barrier is designed to absorb specific macronutrients, vitamins and minerals, electrolytes, and water. The tight arrangement of cells and their protective mucous layer generally exclude bacteria and their breakdown products. This is the mucosal barrier. However, when the cells of the intestinal lining are heated to approximately 42°C, the barrier breaks down because of decreased blood flow to the gut and direct heat injury to the cells. This allows LPS to enter the capillaries and be carried to the liver by the portal circulation and then throughout the body.

When LPS stimulates the production of TNF and IL-1 by immune cells, these cytokines orchestrate a defensive reaction known as the acute phase response. This response is essential for survival, but in conjunction with overwhelming exposure to LPS or large numbers of microbes, it contributes to terrible devastation in the body. Some of the most serious adverse effects of LPS include collapse of the cardiovascular system, activation of the coagulation system, destruction of blood platelets, organ damage from the products of activated white blood cells, and compromise in the integrity of the capillaries causing leakage of plasma into the lungs and other organs. This condition may lead to refractory hypotension, or septic shock, which kills 40% of its victims despite treatment with antibiotics. This is approximately the same mortality rate as that of severe heatstroke, which often causes death even if body temperature has been returned to normal. Thus, prevention of heatstroke is an important public health concern.

There is still much to learn about the pathogenesis of heatstroke and adaptation to heat, particularly at the cellular level. It is clear, however, that most human cells tolerate temperatures above 42°C poorly, and when specific cell types fail, the physiological manifestations relate directly to breakdown in the integrated functions they were designed to support. The near simultaneous heat-related failure of multiple homeostatic mechanisms leads to the complex pathophysiology of heatstroke. The higher the temperature, the greater the number of homeostatic processes that fail and the higher the mortality of the condition.

# Endless Oceans of Sand

Despite the tropical origin of human beings, the hot, dry deserts of the world are no place for ill-prepared adventurers. Through the centuries thousands of ill-fated explorers have underestimated the amount of water required for long desert treks. This miscalculation has led to many deaths by dehydration. Slightly more fortunate sojourners have stumbled across the hot sands for days only to find water in the nick of time or to be saved from certain death by hardy rescuers. The environment of the arid desert is indeed formidable.

The world's largest desert, the Sahara, covers one-third of the African continent, an area the size of the United States (see Figure 8.1). Its temperatures range from 130°F during the day to 20°F at night. Rainfall averages only three to five inches and in the driest areas may be as little as one-sixteenth of an inch a year. Desertification of the Sahara has been one of the most dramatic climatic events since the end of the last Ice Age, some 10,000 years ago. As recently as 3000 B.C.E. much of the region was still arable, but the size of the Sahara continues to expand today.

Failure to carry sufficient water into deserts is an obvious mistake, but water is so scarce that loss of transportation can be equally perilous. The breakdown of one's only vehicle in the remote Sahara may be lethal because the amount of water that can be carried is not sufficient to walk out of the desert (see Table 8.1) or to the next oasis, which may be 100 miles away. This problem reemphasizes the

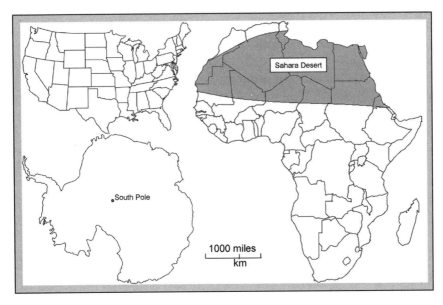

**Figure 8.1.** Outline maps showing relative sizes of Antarctica and the Sahara, the world's two largest deserts, compared to the continental United States.

principles for avoiding the consequences of double failures. For centuries nomads of the Sahara have circumvented this danger by wending their way across the endless expanses of sand between oases in great caravans. The desert is so immense that caravans may travel for days between oases, and for thousands of years the camel has provided the necessary redundancy for desert survival.

## The Camel and the Berber

The foundations of trans-Saharan travel are sometimes attributed to the Libyans who, according to the Greek historian Herodotus, hunted in the desert on horse-drawn chariots. This account has become associated with Moroccan rock paintings that depict horse chariots, which, according to local lore, also carried gold and ivory from West Africa to markets in Carthage and Rome. Although ancient horse charioteers are often credited with Saharan crossings, the early trade routes were probably established by Berber nomads who, at the end of the Roman Empire, frequently crossed the desert with large flocks of camels. Logic dictates that regular trans-Saharan trade awaited the arrival of the camel, for horses cannot live under such harsh conditions. As a desert beast of burden, the camel was unrivaled and enabled transport of not just merchandise but food and water for prolonged desert crossings while the trader traveled on foot.

**Table 8.1.** Minimum Water Requirements for Walking at Different Temperatures*

| TEMPERATURE | MILES | | | | |
|---|---|---|---|---|---|
| °C (°F) | 10 | 25 | 50 | 75 | 100 |
| 26.7 (80) | — | — | 4 | 8 | 13 |
| 32.2 (90) | — | 2 | 8 | 15 | 21 |
| 37.8 (100) | — | 6 | 17 | 28 | 39 |
| 43.3 (110) | 1 | 8 | 21 | 34 | 48 |
| 48.9 (120) | 2 | 13 | 30 | 47 | 65 |

*Based on a walking speed of 3 miles per hour. Water requirements in liters (0.946 L = 1 U.S. quart).

Data compiled from multiple sources.

The camelids evolved in North America and during the Pleistocene migrated across the Bering land bridge to Asia and Arabia, separating them into Old World and New World species. Only Old World camels are suited for desert life. The Arabian dromedary (*Camelus dromedarius*) and the two-humped Asiatic camel (*Camelus bactrianus*), "master" of the Silk Road, can be crossbred to produce fertile progeny. The name *dromedary*, derived from the Greek *dromos* ("road"), conventionally refers to the riding camel. This legendary animal, whose rolling gait has earned it the nickname "ship of the desert," can carry loads of 150 kilograms (330 pounds) comfortably for up to ten days without drinking water (Figure 8.2). The camel was domesticated on the Arabian penninsula 4000 years ago, but it was not widely used in North Africa until perhaps the third century.

As a desert animal, the camel's unique heat and water management strategy is marvelously integrated with its body structure, function, and behavior. Like all mammals, camels are homeotherms, but they are unusual in the extent that their temperature rises during the day and falls at night. When water is plentiful body temperature varies by about 2°C, but with dehydration it can vary by 7°C. By gaining heat during the day, the camel obviates the need to produce liters of sweat. The salt and water so saved preserve the camel's plasma volume. Instead of cooling by sweating, the camel stores heat during the day and dissipates it to the environment at night. In the hours before dawn, when the desert is cold, the camel's body temperature falls to less than 35°C, but by late afternoon it rises to 42°C. The camel does not sweat until its temperature reaches 42°C, when it achieves near-maximum heat storage. Compare this to humans, in whom sweating begins at 25°C and heatstroke mortality is 50% at a temperature of 42.5°C.

The other adaptations of the camel to desert life include powerful measures to conserve water, the ability to regulate brain temperature, and tolerance of extreme dehydration. In addition, the camel's coat limits body heat gain and water loss from the skin. Indeed, the temperature difference between the camel's outer lay-

**Figure 8.2.** The camel and the Sahara. The Arabian camel is not indigenous to the Sahara, having been domesticated and brought to North Africa probably no earlier than the third century. The camel's remarkable desert adaptations, in particular its ability to conserve water, are responsible for opening trans-Sahara trade routes. A camel caravan offers the redundancy of transportation necessary to safeguard the survival of desert nomads. (Copyright Dan Heller, reproduced with permission.)

ers of hair and its reflective skin may be as much as 18°C. The camel's hair illustrates the value of insulation even in the hot desert, for loose reflective coverings slow the rate of heat gain by the body when ambient temperature is higher than is body temperature.

Other water conservation mechanisms of the camel minimize respiratory, urinary, and fecal water losses. Much of what we know of the camel's astonishing suitability to desert life is due to the work of one scientist, Knut Schmidt-Nielsen of Duke University. Schmidt-Neilsen carefully unraveled many of the camel's water conservation mechanisms, including the secret of its remarkable nose (Schmidt-Nielson, 1997).

The camel's nose conserves water by lowering exhaled air temperature and removing water from it. During heat exchange inside the nose, the nasal passages are alternately cooled by inhaled air and warmed by exhaled air. The loss of heat from exhaled air to the nasal passages condenses water much as one's breath condenses on a cold day, and this moisture is retained in the nose. In addition, the vascular anatomy of the camel's nose and head provides a countercurrent heat exchanger that protects the brain from overheating. A camel's brain may remain more than 2°C cooler than its body.

Like many other desert mammals, the camel's kidneys can produce highly concentrated urine. During dehydration camel urine is like syrup and more than twice as salty as seawater (see Table 6.2). Besides retaining more water, this renal concentrating power allows the camel to drink salty water as well as to eat desert plants of high salt content. The urine also cools the camel in hot weather, when the animal wets down its hindquarters and the liquid evaporates. Camel feces contain very little water, and the solid dung pellets are so dry they can be collected by hand and used immediately for fuel.

The notion that the camel's hump stores water was dispelled long ago, as the hump consists almost exclusively of fat, which supports the animal through periods of both starvation and dehydration. Although oxidation of the fat provides some metabolic water, the water stored in the alimentary tract of a well-hydrated camel can amount to 10% to 20% of body weight and is available to forestall dehydration. Over a week or two, a camel can lose more than a third of its total body water without suffering serious consequences, while half of that will incapacitate a human being.

During dehydration the camel's blood osmolarity and viscosity increase radically, and its plasma sodium may reach 160 milliequivalents per liter. The animals can rehydrate very quickly, by about 60% after the first drink and completely in 48 hours; a 500-kilogram camel with a large water deficit can drink 100 liters in ten minutes. When a camel replenishes its losses, rapid absorption of water lowers plasma osmolarity and causes the erythrocytes to swell. If osmolarity falls too low, all erythrocytes lyse, but camel erythrocytes show less osmotic fragility than do those of other mammals. Camel erythrocytes have a robust internal cytoskeleton and an oval shape that looks more like those of birds than of other mammals. This allows them to survive more severe osmotic changes than can the erythrocytes of any other species and enables the camel to drink vast amounts of water rapidly without suffering hemolysis.

The camel's remarkable physiological adaptations to desert life are nevertheless not enough to ensure its survival in extreme heat, although it exhibits a number of behavioral adaptations that also improve its tolerance of heat. The camel couches facing the sun, which reduces the body surface area exposed to the radiant energy and shades the ground under its body. It rests on its knees, which are protected by thick leathery pads, thereby minimizing contact with the hot ground and allowing air to circulate beneath the body. Such heat escape activities are discussed in more detail later in this chapter.

The camel's earliest associations with the Berber people are lost in antiquity, but the origin of these nomads is also one of the many mysteries of North Africa. The Berbers were early inhabitants of the Atlas Mountains, people of European descent who settled in current-day Morocco, Algeria, and Tunisia. They lived on the southwestern rim of the Sahara and migrated seasonally north and south. Learning of the value of gold to the Roman world, they bartered for it in West Africa with salt from great desert domes. The value of salt was so high that it was worth its weight in gold. Gold was carried north as payment for commodities the Berbers could not produce themselves.

Since the eighth century Saharan trade routes have been the purview of the Taureg people, usually regarded as Berbers pushed into the desert by Arabs who were advancing from the east. Recognized by their traditional dress of long indigo robes, they are known as "blue men of the Sahara" and typically wear a veil over the mouth and nose to maintain a seam of moist air from which to breathe. As the camel caravans and loose fitting blue robes suggest, these nomads have adapted to the Sahara primarily through behavior, as time has been short in an evolutionary sense and evidence of natural selection is not apparent. Indeed, humans must use all the behavioral resources at their disposal to survive in the Earth's great deserts.

## Desert Lessons from Pablo and the Haj

A harrowing desert misadventure may produce a near-death survivor with the bloody sweat and desiccated look of severe exposure and dehydration sometimes known as the "Pablo syndrome." Pablo was a Mexican horseman whose dramatic escapade in the Arizona desert was reported in 1906 by W. J. McGee (see Ladell, 1965). Pablo was lost for eight days in the desert and traveled 35 miles on horseback and 100 miles on foot on two gallons of water under cloudless conditions where the daily temperature was 35°C (95°F) and nighttime temperature was 28°C (83°F). Discovered very near the camp for which he had set out, the delirious man had lost 25% of his body weight. His tongue and skin were black and dessicated, and his facial features had shrunken away. The cracks in his skin no longer bled, and he was deaf to all but loud sounds and blind to all but light and dark. His heart rate was slow, and his pulse could not be detected below his knees or elbows. Nonetheless, Pablo made a remarkable recovery from his ordeal, "soaking up water like a sponge," then drinking and retaining vast amounts of water and other liquids until he had regained his former weight. His vision and hearing eventually returned to normal, but his once jet-black hair turned gray.

What accounts for the remarkable ability of this man to survive for eight days in the desert on a quart of water a day? This is less than one-fourth the required "minimum" water rations for the conditions to which he had been exposed (see Table 8.1). In fact, if the numbers can be believed, Pablo had lost nearly 18 liters (4.7 gallons) of his body fluids in little more than a week. Assuming Pablo's normal TBW was 45 liters, he may have lost 40% of it and lived to tell his story!

In the final analysis, Pablo must have done something very right. First, he must have been well acclimatized to heat. Second, he must have behaved in a way that gave him the maximum opportunity to survive. Had he not been found when he was, McGee would not have been able to publish his extraordinary tale.

The problems Pablo encountered are reproduced on a massive scale today during the annual pilgrimage to Mecca in Arabia. After Ramadan, thousands of Islamic sojourners make the Haj, trekking across the Arabian Desert on foot to Mecca. Not surprisingly, when the pilgrimage occurs in the summer, many people fall victim to the heat. Heat exhaustion and heatstroke are rampant, and as many as 2000 pilgrims die from the effects of heat exposure and dehydration. These individuals deprive themselves of the most important biological defense against heat stress, the behavioral response to thirst and the opportunity to seek shade. They die quickly and in great numbers. Therefore, the next section is devoted to a discussion of strategies for avoiding heat stress syndromes.

## Thermal Stress and Behavior

In 1971 the eminent physiologist J. D. Hardy remarked that changes in behavior, in the long run, are the most efficient way to adapt to stressful change in ambient temperature. Hardy was referring specifically to heat adaptation, but his comment applies equally well to other environments. In the animal kingdom patterns of behavior, both instinctive and learned, are critical to thermal balance and to adaptation to thermal stress. These patterns are observed in both warm- and cold-blooded animals. Indeed, every species capable of locomotion displays behavioral adaptation to thermal stress. This may involve simple actions, such as a snake sunning itself on a rock on a cool morning, or elaborate behaviors, such as construction of burrows or dens by mammals to create special microenvironments suited to their physiological needs.

Among the most fascinating natural thermoregulatory behaviors are those of social insects such as bees, ants, and termites. These insects are able to maintain a nearly constant temperature in their hives or mounds throughout the year. The constancy of these microclimates depends not just on the location and insulation of the habitat, but on the activity of the insects in the colony. When ambient temperature increases, the activity in the hive decreases, which decreases the amount of heat generated by insect metabolism. In fact, many animals decrease their activity in the heat and increase it in the cold, and people who are allowed to choose levels of physical activity in hot or cold environments adjust their workload precisely to body temperature. This behavior serves to avoid both hypothermia and hyperthermia (Rowland, 1996).

Many examples of behavioral adaptation to heat can be found in the animal world. These behaviors are matched to the animal's physiology. Many large animals, such as the hippopotamus and water buffalo, take to a wallow to cool off, while others fan themselves to keep cool, as does the elephant with its ears. Rats and other rodents spread saliva on their coats to compensate for their inability to sweat, and some animals, mostly desert dwellers, move about only at night after it has cooled off. As mentioned above, people, given the option, will choose rest over work when the weather is hot. These behaviors are successful strategies that help prevent heat gain or facilitate heat loss.

Perhaps the most common and successful strategy for maintaining constant body temperature is the search for a comfortable spot. One is reminded of this whenever a dog lies down in the shade of a tree on a summer afternoon or next to a fireplace on a cold winter night. All terrestrial animals move to cooler places when the weather is too hot, whether simply to change position relative to the rays of the sun or to search for shady areas with a more comfortable temperature.

Avoidance of overheating is a powerful motivator, and temperature-dependent behavior is given a high priority by the vertebrate brain relative to other survival activities. For instance, in the heat a bird that stands still, or "freezes," in the presence of a predator will move to a cooler place as soon as its body temperature begins to increase to a dangerous level. Animals place hunger above thermoregulatory behavior until thermal stress reaches a point that it jeopardizes survival. Then thermoregulatory behavior dominates hunger; the animal will eat less and lose weight. Overheated animals, however, always seek water because of the power of thirst and the lethal consequences of becoming dehydrated.

For some animals, curiosity or need for activity supersedes thermoregulatory behavior for a brief time, but never to the point of endangering life. In this respect, humans are different and perhaps unique. People can choose other motivations, such as religion, obligation, or competition, above the physiological need to maintain thermal balance. This type of behavior can lead to severe, disabling thermal stress, including heatstroke, frostbite, and even death. The ability to circumvent instinctive behavior during thermal stress by exerting self-control is a higher cortical function and illustrates one of the great strengths and weaknesses of the human brain.

The instinctive search for a comfortable temperature can be demonstrated in the laboratory in almost every kind of animal. The easiest way to demonstrate this behavior experimentally is with a simple device called a shuttlebox. It contains several, usually four or more, interconnected compartments, each maintained at a different temperature, thereby generating a profile of temperatures. When an animal is placed in a shuttlebox, it moves around for a few minutes to become familiar with the environment and then settles down in the most comfortable compartment. The temperature in the part of the shuttlebox selected by the animal correlates closely with its preferred temperature in its natural environment.

Thermal behavior can also be studied in the laboratory by the classic techniques of operant conditioning. In this case, an animal is taught to obtain a reward by pushing a lever. In the classic conditioning studies of B. F. Skinner familiar to most people, the animal learns to receive a morsel of food as a reward for making an appropriate selection. Behavior in response to changes in temperature can be studied using a thermal reward instead of food. For instance, a rat in a cold cage can be taught to press a lever that activates an infrared warming lamp for a few seconds. The animal will learn to press the lever often enough to maintain the temperature in the cage at a comfortable level. The convention for attaining a warm reward in the cold also applies to receiving a cold reward in the heat. Thus, a rat can be conditioned to operate a fan to supply cool air to a hot cage.

For warm-blooded animals, behavioral adaptation to heat provides two unequivocal advantages. First, it conserves physiological work that accompanies heat-dissipating responses such as panting and sweating. Next, by avoiding these responses the animal also conserves water. Both panting and sweating cool by evaporation, and evaporation means water loss from the body, which increases the prospects of dehydration. Indeed, the negative consequences of dehydration are so great that behaviors to limit the need to dissipate heat have been selected for by evolution. Behavior that minimizes heat accumulation is associated with such a strong survival advantage that many variations are found in the animal kingdom. These behaviors, known as heat-escape activities, are precisely attuned to body temperature.

When animals sweat or pant to keep cool, they make up for the loss of water by drinking, which can also dissipate heat if the water is cool. In theory a 70-kilogram person can drop body temperature approximately 1°C by drinking a liter of water at 0°C, but it is important to note that the main stimulus for drinking is thirst, which is regulated by plasma osmolarity, not by temperature. In strictly physiological terms drinking is thermoregulatory behavior only in that it is "designed" to prevent dehydration. In tropical mammals that rely on sweating to keep cool, including humans, it is moot. There are a few drinking behaviors in the animal kingdom that clearly serve a heat dissipating, or thermolytic, role. The emperor penguin provides a good example of thermolysis, for it ingests ice when it becomes too hot, taking advantage of the ice's low temperature and heat of formation to help it dissipate body heat.

## Importance and Regulation of Heat-Escape Activities

The most intriguing and obvious question about heat-escape activity concerns how such behavior is regulated. In fact, heat-escape activity is coupled so closely to temperature regulation that it is relatively easy to predict what an animal will do simply based on the temperature. The animal's nervous system perceives the extent of thermal discomfort, anticipates the effect of the environment on thermal balance, and modifies its activity appropriately. This set of integrated responses is an expression of the true power of behavioral adaptation in maintaining homeostasis. That this behavior is ancient and neurochemical in nature is not in dispute. However, the specific biochemical responses are poorly understood.

An equally important example of the survival advantage offered both by heat-escape activities as well as general behavioral adaptation to temperature is parental behavior. Most adult vertebrates anticipate and act on the needs of their offspring to stay cool as the environmental temperature fluctuates. For instance, mother birds shade their chicks under a wing to help them stay cool. Some animals lick their young and spread saliva on the skin, which evaporates to keep the offspring cool. The best-known examples, however, are cold-escape behaviors in which the parent "instinctively" protects the young from cold by building a den or sharing its body heat. These parental behaviors are particularly crucial for animals whose nervous systems are immature at birth, including human infants. Newborns are not able to fully recruit appropriate thermoregulatory responses to either heat or cold.

Many animals, including warm-blooded mammals, change body position or posture to help regulate internal temperature. This, too, is a familiar behavior in the cold when animals curl up or crouch to conserve heat and groups of animals, such as puppies and kittens, huddle together for protection from the cold. The same types of postural adjustments are seen in the heat. Retracted postures lessen heat loss to the environment by decreasing the surface area available to exchange heat. They are favored when ambient temperature is much colder than body temperature. Extended postures increase heat loss by increasing the surface area of the body exposed to the environment. They are useful to dissipate heat when the environment is hot but still below the body temperature. However, body extension does not work when ambient temperature is warmer than body temperature, because heat accumulates faster. Hence, physiological strategies, such as sweating or panting, must be called into play.

Postural adjustments are used in both hot and cold climates because position is the only way to regulate the amount of exposed surface area without moving to a completely new environment. Natural postural changes are regulated by the detection of changes in heat flux before the body temperature changes, but the actual sensory mechanisms are unknown. Special postures relative to the position of the sun's rays are employed by many animals, including insects, amphibians, reptiles, birds, and mammals. If the goal is heat escape, then aligning the body parallel to solar radiation will minimize heat accumulation. In hot weather an animal may also crouch or squat and face the sun, thereby limiting the profile it exposes to the direct rays of the sun. If warmth is needed, then aligning the body perpendicular to the sun's rays can increase heat gain severalfold depending on the shape of the animal's body. The most elegant example of this behavior is seen in the resting butterfly, which, when aligned perpendicular to the rays, receives maximum radiant exposure; aligned in parallel it is nearly invisible to the sun.

The message from the animal kingdom is clear: heat escape activities are fundamentally important in biology, and human endeavors that circumvent them are extremely dangerous. For normal adults, this is rarely a problem except when athletic competition, military discipline, or religious obligations are involved. Even these activities contain some element of choice on the part of the participants, and the choices made are clearly influenced by the individual's knowledge of (or lack thereof) the risks involved. Not surprisingly, then, most deaths from heatstroke occur in children and the elderly, in whom appropriate heat-escape behaviors are less likely to be activated or acted on.

# Hypothermia

Bad winters usually bring a greater appreciation of the deleterious effects of cold on the body. Most people have had personal experience with mild cold-related skin injuries such as chilblain, the temporary swelling and tenderness of the face or extremities after exposure to wet and cold conditions for a few hours. Residents of higher latitudes recognize the sharp discomfort of frostnip that comes with exposing various parts of the body to subfreezing temperatures. These effects of cold are usually nothing more than temporary inconveniences. More intense and disabling effects of local cold, such as immersion foot and frostbite, result primarily from prolonged operations in extremely cold weather. The greatest danger to survival in the cold, however, is from the effects of hypothermia.

## The Effects of Extreme Cold on the Extremities

Since World War I soldiers and sailors have been taught to protect themselves against immersion foot, originally known as trench foot, the nonfreezing cold injury of prolonged submersion of the feet in cold water. Soldiers are instructed to keep their socks and boots dry and avoid immersion of their feet in water and deep snow as much as possible. Even so, troops of the U.S. Army in Europe during World War II suffered nearly 5000 cold injuries during the bitterly cold winter of 1944–1945. In military operations nonfreezing cold injuries of the foot are encountered primarily in unprepared infantry under winter battlefield conditions when they have become exhausted, dehydrated, and can no longer take care of themselves.

Similar cold injuries occur in civilians who spend days outdoors in cold weather without proper knowledge of the dangers of wet, tight-fitting footgear. Cold water increases heat loss, constricts blood vessels, and restricts blood flow to the toes. The resulting injuries are painful, debilitating, and may require weeks to heal. They also may lead to long-term or permanent neurological symptoms and, in severe cases, amputation. The same is true of frostbite, freezing of the tissues, most often in the feet, which kills living cells by the formation of ice crystals. Ice formation first occurs outside and then within the cells. Intracellular ice causes cell dehydration, cell shrinkage, and abnormalities in electrolyte composition, which denature proteins and enzymes and lead to cell death. Severe frostbite often requires amputation of toes or fingers and may even require amputation of one or more limbs. However, local cold injuries limit human survival primarily by immobilizing the victim and allowing the lethal effects of systemic hypothermia to supervene.

## Settings for Systemic Hypothermia

Survival in the cold is limited primarily by hypothermia, which can be an insidious process and does not require exposure to subzero weather. The dangers of hypothermia are well appreciated due to the publicity surrounding the searches for lost adventurers such as mountain climbers, skiers, and sailors. People lost in the wilderness face protracted exposure to extreme cold, which, despite reasonably good protective thermal clothing, places them at risk for hypothermia.

The dangers of hypothermia are not limited to extreme environments; it occurs in many settings and in all seasons, particularly upon immersion in water. In industrialized countries hypothermia is often encountered in urban settings, where lack of shelter, heat, or protective clothing kills young children and the elderly, who are more susceptible to the effects of cold. The susceptibility to cold of children has a similar basis as susceptibility to heat; children cool faster because they have a greater surface area-to-volume ratio than do adults. The elderly tolerate cold poorly because they lose the ability to generate heat well. The minimum safe indoor temperature for the elderly is 21.1° C (70°F), which is more than adequate for younger adults wearing regular clothing. In the United States alone hypothermia killed or directly contributed to the deaths of 700 or more people each year during the last half of the twentieth century.

The most notorious examples of the ravages of hypothermia occurred during the retreat of Napoleon's army from Moscow in the bitterly cold winter of 1812 and during the German invasion of Russia in World War II. In both cases, the harsh Russian winter was too extreme for the ill-prepared invaders. In September of 1812, 100,000 men of Napoleon's army reached Moscow, which they found abandoned. On the return march food and supplies were scarce, and the inadequately clothed soldiers endured temperatures as low as $-30°C$. On the trek through western Russia, the main force was rejoined by 36,000 rear guard troops. Two months later, 10,000 bedraggled soldiers struggled across the Polish border. The army had lost 126,000 men, or 12 of every 13 members of the original force; most had died of hypothermia.

The insidious nature of hypothermia has often been written about, but even more often the effects of hypothermia are hidden among other more obvious lethal factors. For instance, conventional wisdom about victims of avalanches suggests that they die of trauma or asphyxiation. It is obvious that tremendous bodily damage can be inflicted by tumbling violently down a steep mountain at high velocity and that people who "swim" against the avalanche and fight to keep the snow out of their mouths and noses have higher survival rates. Even so, these people are those who remain conscious and are probably the least severely injured. Three other factors are those who remain important in determining survival after an avalanche: the depth of the snow over the victim, the length of time the victim is buried, and the position of the face (up or down). Death by asphyxia appears to be slightly less likely with the face up because the head melts snow beneath it and a small air pocket forms around the mouth and nose.

Remarkably, over the past 50 years in the United States, no one in direct contact with snow who was buried deeper than seven feet or for more than eight hours has lived. Certainly, the depth of snow must be an important factor, if only because it is impossible to expand the chest under the weight of a tall column of packed snow. That survival time does not expire for nearly eight hours is fascinating, because even slow asphyxia usually causes death within thirty minutes. This suggests a role for body cooling in both the survival and deaths of avalanche victims. One way to explore this possibility is to examine the survival curves of avalanche victims (Fig. 9.1).

In Figure 9.1 panel A indicates that the probability of surviving an avalanche diminishes almost exponentially with respect to time. When the survival curve is fitted to an exponential model (Fig. 9.1B), three phases of the curve are discernible that describe the cumulative risk of death. The interpretation of such curves is complicated and controversial, but interesting hypotheses can be generated from examining them. In this example the initial phase is brief, lasting about an hour, and it meets physiological expectations of asphyxia. The second phase is flatter and ends after about five hours. During this period an immobilized body buried beneath the snow would be expected to cool off. The third and final phase is again rapid and has an even steeper slope than does the initial phase. This final phase is

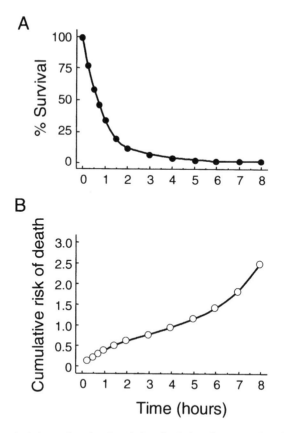

**Figure 9.1.** Survival time of avalanche victims buried under snow. Panel A shows percentage of victims alive at the time body is recovered from a mountain avalanche. Panel B is an exponential fit of the survival curve and suggests more than one cause of death. The early deaths are most likely related to trauma from the avalanche and/or asphyxia that occur shortly after being buried in snow. The steep rise in the cumulative risk of death after five hours suggests an additional factor, most likely hypothermia.

consistent with death from hypothermia. Thus, the exponential survival model suggests that there is more than one reason why people die after being buried in an avalanche, and the curve is consistent with the idea that hypothermia is one of them.

## The Physiology of Hypothermia

A core temperature of less than 35°C (95°F) defines hypothermia. At this temperature the body's defenses against cold, such as shivering and constriction of blood vessels, reach maximal responses. Below this temperature the central nervous system becomes depressed, which impairs memory, decision making, and alertness (Fig. 9.2). Further cooling produces deeper alterations in consciousness, loss of coordination, dehydration (from diuresis), and decreased heat production from shivering. At 30° C (86°F) shivering ceases, and the body becomes poikilothermic; it passively cools to the temperature of the environment. At 27°C to 29°C (80°F to 85°F), depending on the individual, coma occurs, accompanied by absence of voluntary motion and loss of reflexes. The heart slows, and the ventricles become susceptible to arrhythmias, such as fibrillation. Such severely hypothermic individuals no longer have a chance of rescuing themselves but will recover completely if rewarmed properly. In fact, during cardiac surgery, when the body is cooled intentionally to reduce the brain's metabolic rate, recovery is possible from core temperatures less than 10°C (50°F).

For the physician, hypothermia is classified as either primary or secondary. Primary hypothermia is an accident; the victim is exposed to cold that exceeds his or her capacity to defend and maintain body temperature (Danzl and Pozos, 1994). Secondary hypothermia is a complication of a pre-existing disease, such as shock, stroke, burns, severe infections, malnutrition, or ingestion of drugs that impair temperature regulation. Alcohol is important in this respect because it is consumed often in settings where the risk of accidental hypothermia is high. Ethanol impairs thermoregulation in three ways: it interferes with shivering, dilates the blood vessels in

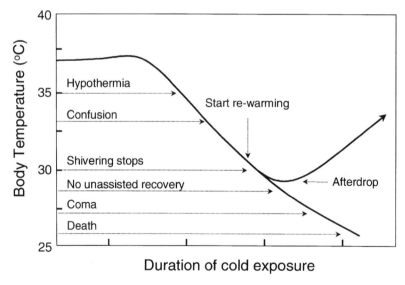

**Figure 9.2.** Graded effects of hypothermia on the human body. The plot shows change in body core temperature (Y-axis) with respect to time (X-axis). The time scale is arbitrary because people cool at different rates depending on ambient temperature, body mass, amount of insulation (fat or clothing), and differences in the extent and efficiency of cold responses.

the skin, and, with chronic use, damages the region of the hypothalamus involved in temperature regulation. The only theoretical benefit of ethanol, as an organic antifreeze to prevent formation of ice crystals in cells, is negligible.

The deleterious effects of hypothermia on the body are a direct result of suppression by cold of cell metabolic processes. As metabolism is suppressed, specific homeostatic functions are gradually lost. For example, sodium and potassium balance are altered due to slowing of the rate of opening of ion channels in the cell membranes. This slows the rate at which excitable cells in the brain and heart, neurons and myocytes, conduct electrical impulses. Cells in various organ systems exhibit different sensitivities to cold so that functions are affected variably. The first signs of hypothermia involve brain function, including impaired memory and decision making and a lower level of consciousness. Although unconsciousness occurs before death, coma is not the cause of death.

The main cause of death from hypothermia is cardiac arrest that results from electrical rhythm disturbance, specifically ventricular fibrillation. The comatose patient can be rescued and rewarmed, although core temperature often continues to drop for a while after rewarming has been started (Fig. 9.1). This "afterdrop" primarily represents continued cooling of areas of great thermal mass after regions of low mass, such as the limbs, have begun to rewarm. Fatal arrhythmias may occur during rewarming because of this. Afterdrop may also be made worse by relief of peripheral vasoconstriction, which allows warmer blood to circulate through the colder parts of the extremities and give up heat to them.

The physiologist's view of hypothermia differs slightly from that of the physician. Physiologists traditionally have studied hypothermia to understand temperature control and to determine how to protect the body from excessive cooling. Therefore, physiologists tend to think about the physical processes involved in cold defense and the integration of responses to it. These processes include the body's attempts to limit heat loss, increase heat production, and coordinate these responses (thermoregulation).

Human beings, as tropical animals, have rather limited defenses against cold. The main defenses against heat loss are skin and subcutaneous fat and the ability to constrict superficial blood vessels of the body. The ability to generate heat is primarily related to the ability to shiver, which is influenced by muscle mass, conditioning, the availability of oxygen, fat and carbohydrates, and endocrine factors such as thyroid and steroid hormones. Thermoregulation requires the endocrine as well as central and peripheral nervous systems to coordinate the body's defenses against cold. Individuals with serious brain and spinal cord injuries or neuromuscular diseases have impaired cold defenses and are at great risk for accidental hypothermia.

A variety of physiological stressors can impair temperature regulation, including dehydration, poor nutrition, fatigue, and sleep deprivation. People at the extremes of age have less robust thermoregulatory responses, and their adaptation to cold is variable. Thus, individual survival time in cold is difficult to predict for two reasons: the physiological response is inconsistent, and the ability to adapt to cold depends on multiple interdependent factors.

Prevention of hypothermia in extremely cold environments requires not only thermal protection but appropriate behavior. Behavioral adaptation is improved with insight into the principles of heat conservation and loss from the body. Knowledge of the effects of wind chill is important. Awareness of the importance of good condition, proper nutrition, and adequate time for adaptation cannot be overstated. Insidious hypothermia can be lethal if clothing allows slow cooling, which interferes with normal physiological responses to cold. In cold weather training soldiers are often taught the practical mnemonic COLD for insulation by clothing: clean, open, layers, and dry. This strategy is designed to avoid water, which conducts heat away from the body twenty-two times faster than does air. Thus, clothing is open during exercise to prevent sweating, and several loose, dry layers improve moisture transfer from the skin and provide better insulation in the cold.

## Unexpected Effects of Cold and Hypothermia

The most extensive experience with extreme cold weather survival comes from polar expeditions. Under the extraordinary conditions of the North and South Poles, adequate thermal protection is paramount, particularly on the Antarctic plateau, where the mean winter (July) temperature averages –60°C (–76°F). In this astonishing cold coffee tossed from a mug freezes before it hits the ground. A bare hand freezes in three minutes, and without exertion for heat generation survival time in simple polar clothing is less than an hour. When properly dressed to leave a South Pole station, the layers necessary to protect the body may include twenty-five pounds of gear. The importance of insulation in such an extreme climate is so critical and obvious that hypothermia can be expected unless the individual is dressed appropriately. In contrast are the subtler effects of cold weather, which affect the survival of thousands of people every year and are overlooked by almost everyone. These effects are also potentially important with respect to human survival.

For decades it has been known that the death rate from heart attacks (myocardial infarction), stroke, and respiratory diseases is greatest in the winter. Many reasons have been postulated, including independent effects of cold exposure, but this has been difficult to prove. Furthermore, excess winter mortality does not correlate well with warm and cold outdoor winter climates, and differences in age, sex, disease prevalence, and respiratory pandemics confound comparisons of cold-related deaths among residents of regions with different climates.

A few unambiguous aspects of the unexpected effects of cold on mortality rate do exist. Half the excess seasonal mortality in winter is accounted for by myocardial infarction and stroke, and their prevalence increases differently as temperature falls in different climatic regions. These deaths are linked, in part, to thrombosis and cardiovascular reflexes activated by cold temperatures. Part of the incidence of thrombosis appears to be due to concentration of the blood. There also may well be more direct effects of cold on activation of the clotting cascade.

Respiratory infections account for half the remaining excess winter mortality. These infections have been attributed to transmission of airborne infectious agents from indoor crowding and longer survival of bacteria in aerosol droplets at cooler air temperatures. Cold also inhibits innate immunity and resistance to infection. Respiratory infections increase deaths from thrombosis, probably by activating coagulation and decreasing clot resolution by inhibiting fibrin clot dissolution after infection.

## The Subtle Effect of Winter on Human Mortality

Whether individuals who protect themselves routinely against the cold have better survival in winter has been an open question for many years. Information is difficult to obtain about whether mortality rates at lower temperatures differ greatly from one climatic region to another. Data on whether variations in cold climate–related mortality correlate with differences in protection against cold were collected in a large European study called Eurowinter published in 1997. Eurowinter attempted to relate winter mortality to the extent of personal protection against indoor and outdoor cold stress for men and women between the ages of 50 and 74 years (Keatinge et al., 1997). The study assessed the increase in winter cold mortality by correcting for age, sex, incidence of influenza, and baseline mortality and relating the winter increases to temperatures in eight European climatic regions. The study also surveyed the extent of protection used indoors and outdoors and related cold mortality to the use of thermal protection.

The findings of the Eurowinter study were rather surprising. The death rate increased more with falling temperature in regions with warmer winters and cooler homes and among people who wore fewer clothes and were less active outdoors. At temperatures less than 18°C (65°F) the increase in winter mortality was greater in warmer than in colder regions. For example, deaths in Greece were eight times higher than in Finland. At an outdoor temperature of 7°C (35°F), average living room temperature was 19.2°C (66.6°F) in Greece and 21.7°C (71.1°F) in Finland. In these two climates 13% and 72% of people, respectively, wore hats when outdoors at 7°C.

A rigorous statistical analysis showed that cold-related mortality was greater with warmer outdoor winter temperatures, colder living rooms, and less bedroom heating. Overall and respiratory mortality were associated with lower percentages of people who wore hats, gloves, and parkas and with inactivity and shivering when outdoors at 7°C. However, deaths from myocardial infarction and stroke were not strongly associated with these environmental factors. This may have been due to insufficient statistical power to detect differences or to local differences in classifying cardiac deaths.

If the Eurowinter findings are correct, several effects of cold on the body could account for the excess deaths. Arterial thrombosis is promoted by dehydration and hemoconcentration induced by cold-mediated urine flow (cold diuresis). However, the association of stroke and coronary thrombosis with cold in this study was weak. Suppression of immune responses by stress hormones during cold exposure may reduce resistance to respiratory infection, and direct effects of cooling on the respiratory tract, such as constriction of the bronchi, facilitate infection. Systemic responses to respiratory infection further increase the risk of thrombosis.

The association between death and inadequate protection against cold suggests that winter mortality may be reduced by improving protection against cold, particularly in climates where the need to avoid the cold is less evident. Evidence of excess winter mortality in people with inadequate indoor heat and outdoor protection implies adverse effects of both outdoor and indoor cold on mortality but does not support the popular myth that a warm house is harmful because the cold challenge is greater when people go outdoors. The practice of wearing a hat is noteworthy because the blood vessels of the scalp do not constrict in the cold. Thus, the head is a particularly high heat loss area.

Eurowinter has been criticized for its ecological design, from which cause and effect cannot be determined because it is not known whether those who died had actually been exposed to greater cold. The study is thought provoking, but the observations need confirmation. Socioeconomic data are also important because low socioeconomic resources in different regions may contribute to excess winter mortality independently of cold exposure. Cold intolerance could reflect poverty and malnutrition as well as or instead of lack of awareness of the dangers of cold exposure. In this context behavioral adaptation depends on how the problem is framed because context influences one's decision about guarding against the perceived stress. The perception of severity of weather in different regions may affect climate-related behavior and ultimately survival or death. The argument is analogous to the choices people make in undergoing extreme heat stress pointed out in the previous chapter.

This discussion raises a final point about cold exposure and survival concerning interactions between food and cold. To generate heat either by shivering or nonshivering mechanisms requires energy, much of which is provided by mitochondria in skeletal muscles that are under the influence of thyroid hormone. Thus, nutritional state plays an important role on one's ability to stay warm. The point is illustrated by the close relationship between hypothermia and specific nutritional deficiencies (Lukaski and Smith, 1996). For instance, thiamine (Vitamin $B_1$) deficiency, common in alcoholism, decreases body temperature and heat production. Thiamine is a critical cofactor for the synthesis of certain neurotransmitters, including serotonin, that are involved in body temperature regulation. Thiamine is also necessary for mitochondria to utilize carbon substrates to produce heat. Classic thiamine deficiency, Wernicke-Korsakoff syndrome, is associated with both hypothermia and impaired temperature regulation.

Deficiencies of iron, copper, iodine, or zinc also independently impair heat production and temperature regulation by affecting various aspects of thyroid hormone synthesis and action. For instance, zinc deficiency decreases synthesis of precursors of thyroid-releasing hormone (TRH) in the hypothalamus, while iron and copper deficiencies decrease thyroid hormone (T4) synthesis. Iron deficiency adversely affects temperature regulation in women and children, whose heat production is less than would be expected in the cold. Normal heat production in iron-deficient women is restored after iron therapy (Fig. 9.3). The critical importance of adequate nutrition for generating heat and for defense against cold is emphasized by the stories of survival on the crystal desert of Antarctica in the following chapter.

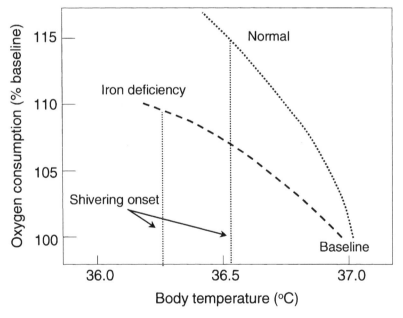

Figure 9.3. Effect of iron deficiency on body heat production. Plot of metabolic response to cold (Y-axis, body $O_2$ consumption) as a function of body core temperature (X-axis). Iron deficiency decreases the metabolic responses to cold. The position of the curve can be restored by repletion of iron stores. (Adapted from Lukaski, Henry C. and Scott M. Smith. 1996. Effects of altered vitamin and mineral nutritional status on temperature regulation and thermogenesis in the cold. In: *Environmental Physiology*. Edited by Melvin J. Fregly and Clark M. Blatteis. New York: Oxford University Press, pp. 1437–1455.)

# 10

# Life and Death on the Crystal Desert

The Earth's most forbidding terrestrial climate, the southernmost continent of Antarctica, is so cold that its winter temperatures do not differ substantially from those on the surface of Mars. On the Antarctic plateau, the elevation averages 2300 meters (7500 feet), and gravitational katabatic winds whip strongly across the ice. The barometric pressure is lower than expected from the altitude on the plateau; effective altitude is 300 meters higher than true altitude and averages almost 2700 meters (9000 feet). Some regions receive less than two inches of snowfall per year, yet the ice sheet averages more than a mile (2 km) deep and pushes the land below sea level. Antarctica is the highest, coldest, driest, and windiest continent. It is also the world's fifth-largest continent, larger than Europe and Australia and one-and-a-half times the area of the United States (see Fig. 8.1). Most of the continent is technically a desert, even though it contains 70% of the fresh water and 90% of the ice on Earth.

## Life in Antarctica

Because most of Antarctica's surface is covered by ice, only about 2% is available for the growth of vegetation. There are no trees or shrubs and only two flowering plants: Antarctic hair grass and pearlwort. Antarctic floras consist mainly of lower plants, mosses, liverworts, lichens, and fungi specially adapted to resist cold and dehydration. These consist of approximately 100 species of mosses, 25 liverworts, 400 lichens, and 20 fungi.

Not surprisingly, the lack of vegetation in Antarctica reflects the reality that few terrestrial vertebrates can survive there. Indigenous vertebrates are limited to birds, primarily on sub-Antarctic islands, including several types of penguins, albatrosses, gulls, and petrels. The continent has no naturally occurring reptiles or amphibians. A modest number of insect and arthropod species are represented, but they live in numbers only on ice-free land at the edges of the continent and on a few rocky locations in eastern Antarctica. On the high plateau, terrestrial fauna are very limited and consist primarily of nematodes. The icy soil is the least diverse ecosystem on Earth.

The cold seas that surround Antarctica are fascinating, and their biology tells much about its unique coastal ecology. Several types of seals, sea lions, and cetaceans inhabit the southern ocean. The waters beneath the coastal ice shelves are a rich breeding ground for krill, a staple food of whales, seals, and sea lions. Among the indigenous Antarctic fish are two remarkable species, the icefish and the nototheniids. Icefish have no hemoglobin in their blood; $O_2$ extracted by the gills is simply dissolved in the plasma and transported to tissues. The blood of nototheniids contains large quantities of glycoproteins, which serve as an antifreeze. Many of these fish are so highly adapted to cold water that they die when the temperature rises above 5°C (41°F). Indeed, the exquisite adaptation of marine life to the waters of the southern ocean speaks to the austerity of the south polar climate.

## The Race for the South Pole

The history of Antarctic exploration is one of the great human epics. Before its discovery pre-Renaissance mapmakers often drew a mysterious southern land, Terra Australis Incognita, in the exact location of Antarctica. The Antarctic Circle was first crossed by James Cook in 1773, but land was not discovered until 1820. In 1838 it was established that Antarctica was a continent and not just a collection of islands, but the continent remained unexplored until the early twentieth century.

The history of the early Antarctic explorers, men seeking distinction by being first to reach the South Pole, is filled with tales of great heroism and great hardship. Misery and death from cold, starvation, and exhaustion were facts of life. The reasons for the successes or failures of valiant explorers, such as Roald Admundsen, Robert F. Scott, and Ernest Shackleton, have been attributed to combinations of experience, determination, and luck. Biologists admit luck only in the absence of other explanations, which means excluding biological factors as determinants of outcome. Because some men died and some survived, biology played some sort of role, even if only in their final hours. What really accounted for the differences in outcome of some of these incredible expeditions?

Just after the turn of the twentieth century, the indomitable British explorer Ernest Shackleton twice tried and failed to reach the pole. Shackleton and his crew survived both encounters, in 1909 and 1914 (Heacox, 1999). The story of his ill-fated expedition to cross the continent, when his ship, *Endurance*, became ice-bound, and the two terrible years spent by his crew in 1914–1916 has gone down in history as one of the most extraordinary tales of human survival. Every one of the twenty-eight-man crew lived through the ordeal, even though *Endurance* was crushed by the ice and sank in October 1915 and the crew was not rescued until May 1916 (Figure 10.1).

**Figure 10.1.** *Endurance* marooned. Photograph by Australian Frank Hurley of Ernest Shackleton's 1915 expedition to the South Pole. The ship was beset by ice and finally abandoned after ten months on October 27, 1915. *Endurance* sank in early November, and the men were forced to live on the ice. (Royal Geographical Society, London, reproduced with permission)

Much has been written of Shackleton's resolute spirit and outstanding leadership abilities, but he was also highly experienced and well schooled in the contingencies of polar exploration. He had been a member of Robert Scott's failed *Discovery* expedition of 1901–1903, after which Scott assigned blame to young Shackleton for their failure to reach the pole. Indeed, Scott was an iconoclast whose expeditions were famous for military discipline and austere provisioning. Shackleton had suffered from scurvy and nearly died of dysentery while traveling with Scott. He learned invaluable lessons from this experience both in terms of how to provision an expedition and how to lead one.

Shackleton's expeditions were robust with food and supplies, and his detractors accused him of extravagance when laying up provisions. However, he understood more of human nature and nutrition than did any other explorer of his time. When *Endurance* sank in October 1915 and the men had to live in tents on the ice, food was plentiful at first but became scarce as the months passed. Shackleton became a resourceful forager of the ice floes, harvesting birds and seals for food. The "Boss" also knew when to strike for help, and his remarkable 800-mile open-sea journey in the spring of 1916 from Elephant Island across the southern ocean to South Georgia with five companions in a twenty-three-foot open lifeboat, the *James Caird*, is almost as astonishing as the rescue of his entire crew.

The race for the South Pole reached its climax in the winter of 1911–1912. The world had expected Scott, leader of the British Terra Nova Expedition (1910–1913), to reach the pole on that, his second attempt. As Scott set the stage, the stalwart Norwegian Roald Amundsen landed at the Bay of Whales in the Ross Sea, about 400 miles from Scott's base and some 60 miles closer to the pole. Amundsen's team of superb skiers and dog-sled handlers, seasoned in the Arctic, set out on October 19 and arrived at the South Pole on December 14, only fifty-seven days later. They planted the Norwegian flag, left letters for Scott, and returned to the Bay of Whales on January 25, 1912, after a 1400-mile journey.

Meanwhile, Scott had chosen four men (Evans, Oates, Wilson, and Bowers) to try for the Pole. They left Ross Island on November 1, 1911. Other members of the party helped set out supply depots some seventy miles apart along the Beardmore Glacier route set by Shackleton in 1908–1909, each stocked with food and fuel for a week. Scott's team began the trip with ponies pulling their supplies, but after five weeks all the over-matched animals had died or had been shot to end their suffering and butchered for meat. The five men ended up hauling the 200-pound sledges themselves to elevations as high as 10,500 feet. They arrived at the South Pole on January 17, 1912, after seventy-eight days only to discover the Norwegian flag left by Amundsen and his party.

Discouraged and beset by bad weather, Scott and his men ran short of food and fuel on the 800-mile (1300-km) return trip. Evans died on February 17 at the foot of Beardmore Glacier. In the three weeks spent crossing the Ross Ice Shelf on the last 400-mile leg of the return journey, the group encountered daily temperatures of −34°F to −43°F. On March 17 Oates walked off into a blizzard to his death in the hope of saving his companions, who were in better condition. The others ran out of food anyway, and Scott's frostbitten foot hampered his ability to walk. Scott, Wilson, and Bowers pitched their final camp eleven miles from One Ton Depot on the Ross Ice Shelf (about 150 miles from their base camp) but were snowed in for ten days before Scott's last journal entry on March 29, 1912. The brave men died in the tent, and their bodies were not discovered until November.

Scott felt he had let his companions down and in his final days wrote that the expedition's misfortune was not due to poor planning but to bad weather and bad luck. "It was no one's fault . . . every detail of our food supplies, clothing and depots . . . worked out to perfection . . . We have missed getting through by a narrow margin which was justifiably within the risk of such a journey."

Scott and his two companions perished a day's walk from a resupply depot and 150 miles from their final destination after having completed more than 90% of the journey. What caused these men to die so close to their goal? Some have claimed it was disappointment at having been beaten by Amundsen. Others point out Scott's decision to use ponies instead of dogs, which were poorly suited to the conditions. Also the weather was unusually cold for March, yet these men would have been well habituated to cold after months on the Antarctic plateau. However, continuous exposure to this kind of cold extracts a colossal toll on the human body, the expenditure of metabolic energy, which can be replenished only by proper nutrition and adequate calories. In addition, to avoid dehydration respiratory water losses at high altitude must be replaced, which requires drinking water derived from melting ice. Without a source of fuel, extremely cold ice would have had to be eaten, which drains precious body heat.

Shivering is a considerable metabolic problem if body temperature falls just 1°C or 2°C, even in acclimated individuals. Although it produces vital heat, shivering consumes large amounts of energy. Heat produced by shivering can be replaced with the heat of exercise, which will keep the body warm, but the energy cost is the same. The cost was increased by the loss of the pack animals, which required the men to pull their sledges. After many weeks of such strain, physical and mental exhaustion set in. To paraphrase Shackleton, being bone tired in this kind of cold, particularly when hungry, makes for a keen desire to lie down and sleep, and to sleep under such conditions is to die.

Exposure to Antarctic conditions requires the consumption of a huge number of calories each day. Voluntary food intake of people carrying out physical activities in the cold varies inversely with temperature and at −40°C (−40°F) amounts to roughly 6000 kilocalories per day. The strenuous expeditionary activities of the kind undertaken by Amundsen and Scott, involving skiing with heavy loads and hauling sledges in the cold, may burn up to 7000 calories a day. Thus, a survival time of eight weeks without food, an average life expectancy, could easily be cut in half by the rigors of Antarctica's cold and altitude. Scott's plan had allowed about 4500 calories per day for each man in his party, and on the return leg they ran out of supplies several times before reaching the next cache. These men lacked adequate rations and continually suffered from hunger.

The failure of such experienced explorers to survive is no mystery when the provisioning is assessed by modern nutritional standards. The provisions set aside for the entire expedition would have sufficed for about ninety days had the party not been beset by unusually cold weather. Assuming a daily requirement of 4500 calories for the first five weeks of the trip, before the ponies were lost, 7000 calories for the remaining six weeks of the trip to the pole, and thanks to hauling lighter loads, 6000 calories on the return, it is clear that the men developed a large calorie shortfall. If, after the ponies were gone and depending on the amount of horsemeat they ate, the shortfall averaged 1200 to 1600 calories per day, Scott and his men would have lost almost 40% of their body weight by the end of March (Fig. 10.2). Even so, three men lived five months, but starvation would have required extensive muscle catabolism. On a return from the pole in very cold weather, the physical work necessary for the emaciated men to drag sledges 400 miles across a mile high ice shelf would have been all but impossible.

## Failure to Adapt to Antarctic Conditions

The unrelenting harshness of Antarctic conditions and the lethal and near-lethal stress faced by early explorers stand in striking contrast to the lack of adaptation to cold in

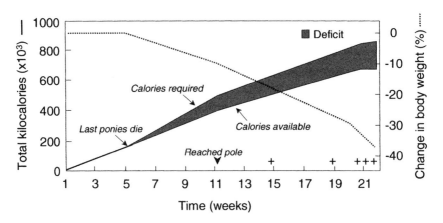

**Figure 10.2.** Plot of energy expenditure for members of the 1912 ill-fated Scott expedition to the South Pole. Estimates of the total number of calories required and those available (Y-axis) are plotted against the duration of the expedition (X-axis) for one man. The average calorie deficit per man for the expedition was probably 200,000 to 225,000 kilocalories. Deaths are shown by + signs. The dashed line shows the estimated decline in body mass during the expedition.

people who today remain at polar research stations for months at a time. Clearly, prolonged stints at polar stations engender unique stressors and share similarities with other isolated and confining environments, such as nuclear submarines and spacecraft. These environments create fascinating and important problems in psychosocial and group dynamics that must be accounted for when selecting crews and planning and managing expeditions. Psychological stressors clearly interact with physiological stressors. However, the physiological stress of confined environments such as a polar station has two components of potential import for human survival.

The first is the surprising lack of evidence that cold tolerance improves in individuals who "winter over" in the Antarctic. This is not true of traditional expeditionary polar explorers, who do show adaptive responses to cold, but it is the case for scientists and engineers who live inside stations. Wintering over in Antarctica today rarely produces more than mild blunting of the shivering and vasoconstrictor responses to cold. On the other hand, these responses themselves may be considered remarkable because exposures to the extreme cold occur intermittently and briefly. The surprise is that briefly exposing small areas of the body, such as the face and hands, periodically to extreme cold produces any adaptation at all.

Prolonged Antarctic residence has been associated with other minor physiological changes, such as the polar T3 syndrome. The syndrome is characterized by altered thyroid function, including an increase in thyrotropin (TSH) release from the pituitary during the winter months, a decrease in circulating thyroid hormone (T3 and T4) levels, and an increase in T3 distribution and clearance by the body. The syndrome is associated with increased energy consumption and stable weight. Its significance is unknown and may be related to cold habituation or adjustment to the pronounced changes in the daily photoperiod at the Earth's poles. In any event, it offers no meaningful survival protection from the extreme climate.

The second point is that Antarctica is a complex environment. It is both cold and high and thus produces conflicting adaptive signals in the form of negative cross-acclimation. The effects of physical condition, quality of nutrition, and changes in circadian rhythms and stress hormones are also important variables that influence human adaptation to complex environments. Therefore, the adaptive responses of outdoor South Polar explorers include some rather surprising and perhaps counterintuitive effects. Most notable are an increase in lipid metabolism and a decrease in body fat content, perhaps reflecting improved physical conditioning or changes in substrate utilization with exercise at high altitude as much as adaptation to extreme cold.

As noted earlier, the high altitude and polar low pressure on the Antarctic plateau can be expected to produce a certain amount of negative cross acclimation with cold. Human acclimatization to altitude is far more robust than is adaptation to cold; in many individuals the latter amounts only to habituation rather than to major physiological acclimation. Although logical, no studies have formally tested the hypothesis that the overriding human physiological response to life on the

Antarctic plateau today is adaptation to altitude. Indeed, the actual altitude at the geographic South Pole is 2,835 meters (9,300 feet), and the effective altitude is 3,200 meters (10,500 feet), and acute mountain sickness is encountered there.

## Engineering Out the Need to Tolerate Cold

The relative lack of cold acclimation in individuals who winter over on the ice reflects, more than anything else, an engineering philosophy. The station is a life-support system designed to provide indoor conditions that make it easy to maintain thermal homeostasis. Hence, the opportunity for cold adaptation is limited primarily to brief outdoor excursions in protective clothing. Such brief and intermittent cold exposures are not conducive to vigorous physiological adaptation, and modern protective air-filled boots and layered clothing provide such good insulation that the body's responses to cold, such as shivering, are limited.

The engineering philosophy behind the design of Antarctic stations requires no defense. Antarctica is the coldest place on Earth, and it is the one continent to which humans have never migrated permanently. Not only is the temperature far below the lower limit of normal human habitats, it is below that necessary to support plants and animals that provide sources of food for people. The scarcity of plants and animals and the 800 miles of open sea to the nearest stable food supply mean that the climate of Antarctica is not one to which humans can be expected to adapt. Furthermore, humans, as a species of tropical origins, do not possess the physiological means of adapting to such intense cold.

The uninhabitability of the South Pole does not hold true for the Arctic region, where hunter–gatherer groups such as the Inuit have lived for nearly 5000 years. The Arctic, too, is bitterly cold, but it has a mean winter temperature of –20°C, which is warmer than Antarctica by about 40°C (72°F). What is it about the difference in temperature between the North and South Poles that is so critical to human survival? A scientifically rigorous answer to this question has not been found, and testing hypotheses is not simple.

An important and perhaps major part of the answer clearly has to do with the availability of food in the Arctic, whereas none is available in the Antarctic. On average the Arctic is capable of supporting one person for every 150 square miles of land. One can also assume from the cooling times of the human body at different temperatures that the degree of cold is very important. Body-cooling rates may differ by nearly an order of magnitude under the two climatic extremes, which means that a potentially lethal accidental exposure to the cold must be corrected ten times more quickly in the Antarctic than in the Arctic. Thus, Antarctica is far less forgiving of the natural human proclivity to make mistakes.

The Antarctic plateau is so remarkably inhospitable to terrestrial vertebrates that it has not even allowed survival by the evolution of cold adaptive metabolism such as nonshivering thermogenesis and hibernation. The difference in temperature and altitude between the semi-inhabitable peninsula and the high plateau is too great to have allowed migration and natural selection of anything more advanced than nematodes. The implication is that the collective stresses of cold, high altitude, and food scarcity stretch the limits of terrestrial vertebrate physiology too far to permit them to live long enough to acclimatize and multiply.

## Human Acclimation to Cold

How do people acclimate to cold, and how does human cold tolerance compare with respect to other mammals? The answers require an understanding of three factors: heat generation, body insulation, and cold habituation. Humans improve on all three parameters during chronic cold exposure, but their responses are unimpressive compared to real cold-weather mammals. At best, these responses do not even amount to the protection provided by a regular business suit. The insulation of a business suit is approximately one standard unit, called a clo. The insulation of a clo is usually given in units of heat flow: area × temperature ÷ by the power, or $(0.155 \text{ m}^2 \times °C)/\text{watt}$. A physiological adaptive capacity of one clo is indeed tiny, as the blubber and pelts of many cold-weather mammals exceed that by an order of magnitude or more. Thus, modern humans, who migrated out of the tropics only 50,000 years ago, have developed a more efficient and flexible strategy to stay warm: adaptation in the form of clothing. Nonetheless, there is much to be learned about human survival by how the three physiological factors interact in cold adaptation.

Warm-blooded animals generate body heat in three ways. The first is obligatory heat generated by resting basal metabolism. This is the basal metabolic rate (BMR), which in a 70-kilogram adult man is a little more than 1 kilocalorie (kcal) per minute, or roughly 1600 kcal per day. The BMR is usually assigned a value of one metabolic equivalent, or met (1 met=1kcal per kilogram per hour), and to generate 5 kcal, the body consumes about a liter of oxygen. The second way to generate heat is by the mechanisms that regulate body temperature. This regulatory heat takes two forms, but shivering is the more important. Human adults depend entirely on shivering, while infants generate a significant amount of heat by nonshivering thermogenesis (NST). The third type of heat generation is physical work or exercise, which is efficient for heat production because it is inefficient with respect to the work accomplished.

The ability to shiver is a unique characteristic of birds and mammals that is not shared by lower vertebrates. It is another distinguishing physiological feature of homeothermic, compared to poikilothermic, animals. The rhythmic involuntary muscle contractions of shivering consume oxygen and generate heat in a manner similar to exercise. In birds shivering is done mostly with the large pectoral muscles involved in flight; in mammals it occurs primarily in the large muscles of the trunk.

Shivering produces a maximum sustained metabolic rate of about three mets in an adult, which is less than half the median for birds and mammals, for which values above seven mets are possible. The third type of heat generation, physical work or exercise, at its peak can attain 15 mets in trained people. Work is a superior source of heat production; under cold conditions even moderate exercise suppresses shivering.

An obvious way to investigate human cold adaptation is to study permanent residents of polar regions who have made regulatory adjustments to the cold. Somewhat surprisingly, these responses do not appear to improve the ability of the body to protect internal body temperature. The most common adjustment is less vigorous, or blunted, shivering in response to a change in air temperature. For instance, unacclimatized people exposed to air at 5°C (41°F) normally increase metabolism to 1.7 met, whereas Arctic Inuits increase metabolism to only 1.3 met at the same temperature. The Inuit also have less intense cold-induced vasoconstriction than do nonpolar people. Similar observations have been made in nomadic Lapps, who show smaller increases in metabolism and a greater fall in rectal temperature than do control subjects in the same cold environment. This type of cold acclimation, called habituation, is the most common of the three patterns of cold adaptation in humans. How much of the difference in adaptation is related to lifelong cold exposure and how much to genetics is unknown.

Two other patterns of cold adaptation, the metabolic pattern and the insulation pattern, are observed less often than is cold habituation. The metabolic pattern involves greater heat production with more pronounced shivering and perhaps more nonshivering thermogenesis in the cold. The insulation pattern is complex and involves augmentation of vasoconstrictor responses in the skin and superficial muscles and may, but does not necessarily, involve increased fat thickness in the subcutaneous tissues.

Some technologically undeveloped peoples show unusual or exaggerated responses to cold. The reasons are not entirely clear, although many prominent physiologists in the early twentieth century believed that people who had not yet turned to clothing or living in heated homes were more likely to exhibit primal cold adaptation than were multigenerational residents of technologically advanced societies. However, this premise has never been convincingly supported by scientific evidence.

Australian Aborigines live in desert regions where nighttime temperature can reach 0°C. For centuries and until after World War II, the Aborigines were outdoor nomads who wore no clothing and slept on the ground. Aborigines who had spent their lives under these harsh conditions did not shiver at night, their metabolism remained constant, and rectal and skin temperatures fell more in the cold than did those of control subjects. Even so, the conductance of heat away from the Aborigines' body was also lower. They had stronger skin vasoconstrictor responses, which provided them with better-insulated body shells. This meant that their bodies lost heat more slowly, but because they did not shiver, core temperature fell more rapidly. Thus, as in many other mammals, a most advantageous adaptation to cold appears to be to conserve energy rather than to conserve heat.

The cold responses of the Kalahari Bushmen of South Africa also have been studied. These tribesmen do increase their metabolism on cold nights, but significantly less than do control subjects. They show cold habituation in terms of a suppressed shivering response, but they do not appear to have a different skin vasoconstrictor response than do other peoples. They behave more like non-acclimated Westerners than like polar natives such as the Inuits, possibly because they have lived so long in a more temperate climate.

What accounts for three different patterns of cold acclimation in human beings? Clearly, habituation is most important and consistent with the typical cold adaptation of our mammalian ancestors. However, there is no mistaking the existence of the other patterns, and there is overlap in the expression of the three types of cold adaptation. Differences can certainly be attributed to the intensity, duration, and frequency of the periods of cold exposure. One particularly important factor appears to be the rate of body cooling. Slow cooling produces less shivering than does rapid cooling because the rate of temperature change, which governs the intensity of shivering, is less. The rate of cooling may help determine the pattern of adaptation that develops during chronic cold exposure. These principles are illustrated in Figure 10.3.

The ability to develop intense vasoconstriction allows heat conservation and slows internal cooling. Hence, energy is conserved because shivering does not become so intense. This pattern of acclimation often accompanies intense local

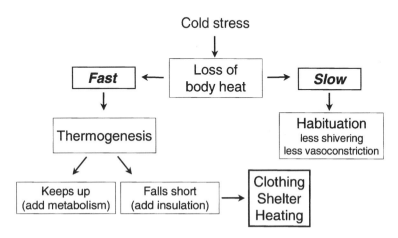

**Figure 10.3.** Patterns of human adaptation to cold. The branching points in the diagram are based on whether the rate of cooling is fast or slow. With fast cooling the body responds by increasing metabolism (e.g. shivering) or by adding insulation. With slow cooling body temperature is allowed to fall with a less vigorous attempt to compensate.

cooling of the face or extremities. The rate of body cooling and the intensity and duration of local cooling probably coordinately regulate the pattern of acclimatization that emerges over several weeks of cold exposure. The bottom line is that the slower the rate of cooling, the more likely are people to show the older phylogenic response of habituation than other patterns of cold adaptation. Regardless of the type of cold exposure, the human adaptive response is rather weak and serves only as a short-term aid to survival. Prolonged exposure to ambient temperature less than 10°C is so poorly tolerated by the human body that this is best defended against by behavioral adaptation to avoid excessive cooling, such as clothing or heating in the home. This engineering strategy, to which humans committed long ago, is for all practical purposes irreversible.

Since the discovery of fire people have regulated the temperatures of their dwellings. Modern homes are actively heated and cooled often to within 3°C year-round. Of the places one might ever care to visit, there are almost no limits to this kind of technology; people have demonstrated an ability to survive in protective suits in the cold (and heat) of outer space and in heat so extreme that it approaches the temperature on the surface of Mercury. The strategy of suit building relies on intelligent behavior to circumvent limit physiology, but doing so also circumvents biological adaptation and quite likely the evolution of physiological changes. It is tempting to believe that humans are gradually replacing certain aspects of physiological evolution with behavioral evolution in the forms of intelligence and technology.

A fascinating theoretical implication of switching from biological to behavioral adaptation concerns the extinction of species. In 1973 Leigh Van Valen proposed the "law of constant extinction" based on survivorship curves generated from taxonomy data. These "curves" are often actually linear, and linearity implies a constant rate of extinction. If true, the probability of extinction is independent of the age of a species. Van Valen's proposal was that situations in which the probability of extinction is nearly constant over millions of years could reflect coevolution, whereby a change in one species, such as a prey, could lead to extinction of another, such as a predator. In this way a species would have to evolve in order to preserve its existence. Van Valen called this idea the "Red Queen hypothesis."

The term *Red Queen hypothesis* comes from a passage in Lewis Carroll's *Through the Looking-Glass* in which Alice encounters the Red Queen, who is leaving the looking glass house for the garden. Alice decides it would be easier to see the garden if she accompanied the Red Queen to the top of the adjacent hill. At the crest the Red Queen begins to run faster and faster. Alice runs, too, but is puzzled to find that neither of them is moving. When they stop, they are in exactly the same place, and the Red Queen explains, "It takes all the running you can do to keep in the same place." The analogy suggests the prospect that evolutionary change may be needed to maintain the fitness of species and that cessation of change may lead to extinction. Although the idea has generated much controversy, it describes an essential dynamic of the coevolution of species. Parallel reasoning raises the possibility that behavioral adaptation, if driven by stimuli different than physiological adaptation, could move humans away from critical coevolutionary environmental factors and lead to our extinction.

## Estivation

There are diverse behavioral alternatives for adapting to extremes of temperature in the animal kingdom, particularly in invertebrates. In warm-blooded terrestrial animals, however, the most striking adaptations are estivation and hibernation. These behaviors are similar in that they involve periodic bouts of controlled torpor in response to seasonal changes in environmental conditions. Estivation occurs primarily in hot, dry conditions, while hibernation is a response to winter cold. The onset of torpor can lead to a reduction in resting metabolic rate by as much as a hundredfold. During hibernation internal body temperature may fall to within 2°C of ambient temperature. Both states represent advanced forms of thermoregulation and require extensive coordination of the physiological systems of the body. Behavior and activity are so closely linked that it is difficult to say whether the behavior evolved as a natural consequence of a physiological adaptation or the physiological adaptation resulted from a change in behavior. In either event, these states illustrate most dramatically the intricate associations possible between behavioral and physiological adaptation.

Estivation is best defined as a pattern of adaptation that restricts metabolic activity and therefore metabolic need in response to harsh conditions (Riedesel and Folk, 1996). A broad definition is necessary because estivation does not operate by a single set of principles. Rather, it produces the same set of results for organisms. It buys time for an individual to survive until conditions become more favorable. Not surprisingly, then, different species have developed different ways to estivate. Estivation has an ancient phylogeny and occurs in all animal orders, including invertebrates. Even ticks estivate until they sense butyric acid in the air, which indicates the presence of a nearby mammal. Unicellular bacteria and fungi that form spores use the same principle to wait for favorable growth conditions. Similar strategies in vertebrates, including mammals, are governed by a range of factors and involve different biological mechanisms. For instance, lack of food and water usually trigger mammalian estivation, whereas lower animals respond to a wider range of environmental cues, including diurnal light and dark cycles.

Estivation in mammals is best exemplified in desert rodents that live in semiarid regions with large diurnal temperature fluctuations and variable, unpredictable rainfall. It is usually controlled by the thermoregulatory system. When ambient temperature declines to 20°C to 25°C (68°F–77°F), body temperature falls to within 1°C or 2°C of ambient temperature. Despite this body cooling and its attendant decrease in metabolism, the animal remains awake and mobile and can forage for food. These periods of torpor commonly last just a few hours, although in some cases they can last as long as several days.

Two other points about estivation are worth noting. First, despite a great deal of study, the torpor mechanism is not understood. In some species metabolism appears to be regulated by an endogenous opiate system in the brain. Endogenous opiate activity is coordinated not only with the range of daily temperatures in the habitat but with the cycle of light and dark, which may play a role in the response. Second, estivation provides a survival advantage to animals that have low fat reserves and small body masses. The high surface area to body mass ratio of estivators is also conducive to entry into and arousal from torpor, both of which may be accomplished in a matter of minutes. In this respect the behavior is intermediate in terms of temperature regulation between cold-blooded vertebrates such as reptiles and more highly regulated warm-blooded animals. From a behavioral perspective, it should also be noted that estivation is an activity that anticipates a difficult period ahead: the animal enters torpor and conserves energy while body weight and energy reserves are still normal.

## Hibernation

Hibernation is a striking seasonal adaptation and the most extreme of the metabolic adaptations to cold by homeothermic animals. Birds do not show true hibernation, but many do show appreciable diurnal variations in body temperature (Roberts, 1996). Some demonstrate controlled, or shallow, hypothermia and allow body temperature in cold weather to fall to about 25°C. A few, such as whippoorwills, hummingbirds, and nightjars, lower body temperatures to 5°C to 10°C. Unlike hibernation, however, these are diurnal, not seasonal, responses. Some large mammals also show exaggerated diurnal temperature variations, but, except for hibernators, body temperature tends to stay within about 10°C of basal temperature. Humans regulate internal body temperature rather tightly and generally defend it vigorously against it falling below 36°C. Realistically, humans show no evidence of ever having been able to hibernate, although there have been long periods in history when it would have been useful.

To appreciate the advantages of hibernation, it is worthwhile to review another thermal adaptation, nonshivering thermogenesis (NST), which is necessary for arousal from hibernation. NST is a highly conserved mechanism of heat generation found almost exclusively in mammals. The primary site of NST is brown adipose tissue (BAT), or brown fat, present in various amounts in mammalian species (Hayward and Lisson, 1992). BAT is not found in birds or nonplacental mammals such as monotremes and marsupials. Thus, BAT is a relatively new evolutionary advance. In placental mammals that use NST, the response can be instigated within minutes of the onset of cold exposure. BAT contributes to heat generation particularly in cold-acclimated mammals, in which, remarkably, it may account for more than a third of heat production. In general, younger, smaller, and hibernating mammals depend more on NST than do older, larger, and nonhibernating mammals. In humans only infants use much BAT; adults rely almost exclusively on shivering to generate heat during cold stress.

BAT is richly innervated and produces heat under the influence of the adrenal stress hormone norepinephrine. Norepinephrine binds to receptors on cell membranes known as alpha- and beta-adrenergic receptors, which, through a coupling, or G, protein, activate the intracellular enzyme adenylate cyclase. Adenylate cyclase uses the high-energy compound adenosine triphosphate (ATP) to generate the intracellular second messenger molecule cyclic adenosine monophosphate (AMP), which signals a number of changes inside the cell, including ion movement, fuel preference (more oxidation of fatty acids), and gene expression. In BAT beta-receptor stimulation causes the cell nucleus to produce RNA that encodes for a protein called uncoupling protein (UCP), or thermogenin. This model is illustrated in Figure 10.4.

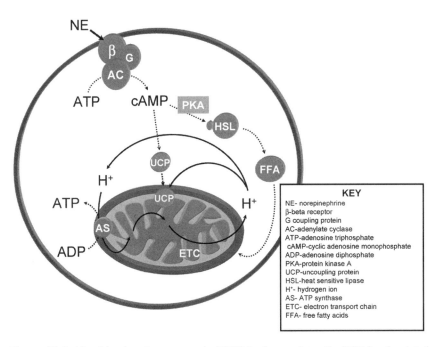

**Figure 10.4.** Nonshivering thermogenesis (NST) by brown fat cells. NST is stimulated by release of the adrenal stress hormone norepinephrine (NE), which leads to the production of uncoupling proteins (UCP) by the cell. UCPs enter the mitochondria and permit dissipation of the proton gradient across the inner membrane. This allows electron transport to proceed without conservation of energy in the form of ATP. The electron transport energy is dissipated as heat.

Thermogenin alters the activity of the cell's furnace, the mitochondria, by inserting itself into the inner membrane in BAT and some muscle cell mitochondria. Thermogenin is activated by norepinephrine in the presence of thyroid hormone and causes mitochondria to transfer electrons from the donor (reduced pyridine nucleotides, NADH) to molecular $O_2$ without conserving chemical energy in the form of ATP. This occurs because thermogenin allows protons ($H^+$) to leak across the inner membrane into mitochondria independently of the enzyme that synthesizes ATP (ATP synthase). The synthase requires the passage of protons ($H^+$) to convert adenosine diphosphate (ADP) to ATP. Thus, the energy of electron transport is liberated as heat and helps keep the body warm.

There is more than one uncoupling protein in mammalian organisms, but only thermogenin (UCP-1), which is uniquely expressed in BAT, appears to be required for cold acclimation (Hanák and Ježek, 2001). This has been demonstrated in mice by deletion of the UCP-1 gene. Several other uncoupling proteins, including UCP-2, UCP-3, UCP-4, and related proteins, have been identified and show a wider tissue distribution. In human adults, who have very little BAT, UCP-2 and UCP-3 are present and have been proposed to play a role in weight regulation and habituation to cold. The primary physiological functions of UCP-2 and UCP-3, however, may not be related to NST, because mice that are genetically deficient in either protein are neither cold intolerant nor fat. Other functions have been suggested because UCP-2 and UCP-3 deficient mice overproduce reactive oxygen species, and UCP-2 deficient mice secrete extra insulin. These UCPs may regulate production of reactive oxygen species by mitochondria, but the purpose is unknown.

NST is often coupled with hibernation as a definitive biological strategy for ensuring survival of an individual under severe natural conditions. Hibernation curtails metabolic activities to match the lack of food in the environment. This metabolic restraint is accomplished by allowing body temperature to cool to within a degree or two of ambient temperature for most of an entire season. Members of six mammalian orders hibernate, including some primates. In these mammals hibernation is a physiologically unique state induced by mechanisms linked to shortening of the daily photoperiod. Before hibernating the animal becomes voracious and feeds constantly to store up large quantities of fat, upon which it will depend almost exclusively during the winter. In small hibernators body weight may double in four or five weeks, while in larger animals, such as the black bear, weight may increase by one-third. A bear may hibernate in its winter den for four to six months, neither eating, drinking, urinating, or defecating.

As the time for hibernation approaches, an animal's activity declines, and sleep intervenes. Entrance into hibernation always begins with sleep, usually a pattern of slow-wave sleep. Although hibernation has been called the *Winterschlaf* ("winter's sleep"), it is much more than an extension of sleep. It represents a distinctive lack of consciousness, neither sleep nor coma, and is one of biology's great remaining mysteries. The animal curls up, breathing slows and eventually becomes intermittent, and body temperature falls rapidly. The heartbeat slows to five or six per minute, and the metabolic rate declines to as little as 1% of the normal basal rate.

In hibernating mammals, body cooling and decline in metabolism go hand in hand, and herein lies a "chicken or egg" paradox. The body cools as metabolism declines, but cold also reduces metabolism, roughly halving the metabolic rate for every 10°C-decline in temperature. This principle, called the Arrhenius relationship, predicts a metabolic rate of only about 3% of the basal value at 0°C, but hibernating metabolism cannot be explained by this effect of cooling alone. For example, the hibernating Arctic ground squirrel maintains a body temperature near 0°C even when the outside temperature is –20°C. The squirrel's metabolic rate falls until ambient temperature and body temperature approach 0°C, but metabolism rises when the outside temperature gets colder and maintains body temperature at 0°C. Thus, the rate of metabolism during hibernation is closely regulated, and its suppression does not simply represent loss of function. There is clear evidence that other physiological processes are also closely regulated during hibernation. The heart does not fibrillate, new protein synthesis is inhibited, antioxidant defenses are increased, immune function is suppressed, and the clotting time of the blood is prolonged (O'Hara et al., 1999).

The periodicity, or chronobiology, of the hibernation cycle is regulated with an intrinsic clock, or oscillator, located deep in the brain. This clock senses the daily photoperiod in the seasonal process of the "circannual" rhythm. The primary cue is light, and all such cues are aptly called *Zeitgebers* ("light givers"). At the onset of hibernation, the electrical activities in different regions of the brain shut down in an orderly sequence. However, a few deep brain regions, such as the suprachiasmatic nucleus (SCN), remain electrically active while all else is silent. Cells in these active regions are involved in regulating the hibernation cycle, but none has been pinpointed as the oscillator. For many years it has been known that the SCN oscillates intrinsically, discovered by observation of variations in its metabolic use of glucose, which is always greater in the daytime. This metabolic rhythm persists under constant lighting conditions, so it is an intrinsic property of the SCN. Indeed, metabolic oscillations of the SCN are present even when the area is removed from the brain and placed in a tissue culture bath. The SCN also appears to contain more than a single oscillator because it can be dissociated into two or more independent components under some conditions. Although generally considered to be the master clock, there are clearly circadian patterns of brain activity that do not require the SCN but appear to be synchronized by it.

In the 1970s, R. Y Moore discovered in mammals direct nerve projections to the SCN of the hypothalamus from the retina. These projections, the retinohypothalamic tract (RHT), led Moore to propose that the SCN regulates circadian rhythms through the eyes, but the retinal photoreceptors appeared not to involve the normal visual system, that is, the image forming rods (Moore, 1997). This concept was supported, in part, by entrainment of normal light rhythms in some blind people. Recently it has been discovered that stimulation of the SCN in response to changes in *Zeitgeber* is mediated by highly specialized cells in the retina of the eye that are distinct from normal visual photoreceptors (Berson et al., 2002). Beyond this little is known about the input to the clock that regulates the timing of hibernation with such remarkable precision.

The presence of circadian rhythms in almost all species suggests they impart a strong survival advantage to animals. In other words, it is to the animal's advantage to repeat certain behaviors at certain times of day, perhaps simply because they have proven successful on previous days. Behavioral predictability means the internal clock must be synchronized to external stimuli in the environment, such as light. Although the circadian clock is also involved in seasonal rhythms by timing the length of the day, endogenous circannual clocks appear to have their own intrinsic behavior.

Entry into hibernation, which also is not well understood, appears to be initiated by the release of specific chemical mediators into the circulation. The first to be described is a plasma factor sometimes referred to as the hibernation induction trigger, or HIT. The existence of HIT was proposed on the basis of experiments in which infusing plasma from hibernating ground squirrels into active squirrels induced summer hibernation. Although the responsible agent has not been identified, it is a protein or peptide that associates with plasma albumin, and its effect varies from species to species.

Endogenous chemicals, such as opiates, melatonin, and testosterone, have been implicated in the onset of hibernation in some species. In 1998 an opiatelike peptide was isolated from the plasma of hibernating woodchucks that plays a role in suppressing metabolism. This peptide has activity as a delta-opiate receptor agonist. Synthetic analogs of the peptide have been found to protect survival of cells from the brains and hearts of other species during hypoxia.

An important behavior of all hibernating animals is periodic arousal from torpor during hibernation. These periods of arousal occur only every few days, but body temperature and metabolism return briefly to normal by a combination of shivering and nonshivering thermogenesis. Neither the mechanisms that control this behavior nor its exact purposes are clear. It is clear that arousal and the increase in activity are associated with substantial energy costs, and thus it must be physiologically important. The most obvious explanations, such as the need to clear metabolic waste, have been investigated and have little to recommend them. A more attractive explanation, in general terms, is that an animal must arouse periodically to restore some critical parameter of intracellular homeostasis.

## Hibernation, Energy Conservation, and Suspended Animation

Hibernation has long fascinated scientists for its potential as a form of suspended animation that might be used to extend human life in the presence of incurable disease, to protect the brain from lack of oxygen, or to conserve energy during prolonged spaceflight. The coordinated adaptations of the hibernating mammal allow its nervous system to achieve tolerance to a range of injuries, such as head injury and oxygen lack. If these protective adaptations can be understood at the cellular level, then novel approaches may be developed to treat stroke, head trauma, or Alzheimer's disease (Drew et al., 2001).

The concept of using hibernation to suspend animation to save energy and avoid monotony in spaceflight has theoretical and practical limitations. The idea has attracted attention because hibernation efficiently conserves energy, and the animal returns to its regular activities in the spring as though nothing extraordinary had happened to it over the winter. The notion that hibernation increases longevity by slowing metabolic rate has a simplistic rationale and no hard scientific support.

The subject of saving energy by hibernation is a fascinating one. During the winter the hibernator depends almost exclusively on fat as a metabolic fuel. Fat is the most energy-efficient biological fuel, and its use has the added benefit of sparing protein. The animal oxidizes stored fat, which can account for half its prehibernation body mass, without wasting valuable muscle protein. This choice of fuel means that when the animal emerges in the spring, its muscle bulk, strength, and mobility are fully preserved. A suitable explanation of how this aspect of hibernation is regulated has not been found, but understanding it could greatly benefit people who are expected to be immobile for long periods of time due to injury or critical illness.

How much energy does an animal actually save by hibernating? The answer to this question is crucial if one wants to explore hibernation for making prolonged human space travel tolerable and efficient. A simple thought experiment to compute maximal hibernation time during space travel requires an estimate of two variables: metabolic rate and the number of calories per day needed for a body in suspended animation on an interplanetary ship. Because humans do not hibernate naturally, the metabolic rate estimate must be extrapolated from that of mammals that do hibernate. Precise estimates of metabolic rate in hibernating mammals are hard to come by because the calculated values are greatly affected by the frequency and duration of arousal periods. In careful field studies using radiotransmitters to record body temperature, energy savings in some species have been computed to be as high as 90%. It is also important to realize that despite this savings, all the energy needed for hibernation was already on board, so to speak, stored in the prehibernation period as fat.

To compute survival time for a hibernating astronaut on a spaceflight, it will be assumed that metabolic rate is reduced by 90% (to 10%) of the basal value, that the astronaut is a man of lean body mass who normally weights 70 kilograms, and that he has eaten enough in the months before the flight to gain 30 kilograms (66 pounds) of fat. The water requirement will be assimilated by metabolic water production. If the 100-kilogram (220-pound) astronaut lies still all day, he would need about 1700 kilocalories per day to meet his needs. Because 1 kilogram of fat is worth approximately 9000 kilocalories, he has stored away 270,000 kilocalories of energy. The 30 kilograms of fat would support him for 159 days. During hibernation his average calorie requirement would fall to 170 kilocalories per day, and he would be able to survive for 1588 days, or more than four years!

If the ship cruised at an average velocity of 90% of light speed, the astronaut could hibernate to within one astronomical unit of the Sun's closest stellar neighbor, Alpha Centauri (which, unfortunately, has no planets). The astronaut then would have to awaken and gain another 30 kilograms of fat, which would require a month and, assuming a balanced diet of 5500 kilocalories per day, 50 kilograms of food before returning to hibernation for the trip home. He would arrive at Earth, neglecting relativistic effects, approximately eight years after he left weighing 70 kilograms. During the round trip he would have spent only one month awake, would have had no exercise, but would have consumed just 50 kilograms (110 pounds) of food. The savings in food alone would amount to roughly 450 kilograms, or almost half a ton. Even more remarkably, however, the astronaut would have realized a savings in $O_2$ of more than a ton (see Chapter 20). Even if the technology existed to accelerate a ship containing a 100-kilogram human, 50 kilograms of food, and 140 kilograms of $O_2$ rapidly to 90% of light speed, such a fanciful trip would not be possible without a great deal more understanding of mammalian biology.

# 11

## Survival in Cold Water

In comparison to other types of cold exposure, immersion in cold water encompasses a special set of biological concerns because of the rapidity with which the human body cools in water. Water has a thermal conductivity 22 times that of air and a heat capacity 3550 times that of air. Compared to marine mammals, such as whales, seals, and walruses, and aquatic Arctic animals, such as the polar bear, the ability of the human body to tolerate immersion in cold water is negligible. This poor tolerance of cold water is directly related to lack of body insulation, such as blubber, with which to retain the metabolic heat of the body. The point is amply illustrated by the appalling history of loss of life in shipwrecks at sea in the northern and southern latitudes of the world.

### The Sinking of the *Titanic*

On Sunday night, April 14, 1912, the British liner *Titanic* struck an iceberg in the North Atlantic and sank in two and a half hours. The seas were calm and the sky clear at latitude 41° 46' N that night, but the water temperature was less than 0°C. Of the approximately 2207 people on board, only 712 were able to enter lifeboats. Although some passengers and crew members went down with the ship, perhaps 1000 people entered the freezing water wearing regular clothing and life jackets.

The mighty liner broke in half and slipped beneath the surface at approximately 2:20 A.M. By 3:00 A.M., according to the accounts of the people in the lifeboats, the cries for help of those in the water had completely died out. The final outcome of the tragedy is well-known to every school child in the world. All 712 people survived in the boats, but of the thousand who entered the water and remained there for more than a few minutes, only the chief baker was recovered alive. The baker, a particularly corpulent man, floated in his life jacket in freezing water for almost two hours before being pulled aboard one of the lifeboats. Thirty men lived through the night by climbing out of the water and clinging to the hull of an overturned lifeboat.

Within three hours of receiving the distress call and two hours after the *Titanic* went down, the Cunard liner *Carpathia* managed to steam 58 miles to the scene of the disaster and retrieve the first boatload of survivors. Within six hours of the sinking, all the lifeboats had been recovered, but the crew of the *Carpathia* was unable to find a single survivor in the water. Over the next two weeks the bodies of some 300 victims were located and pulled from the sea by other ships. The dead were floating upright in their life jackets, faces safely out of the water, many wearing peaceful expressions as though merely asleep. They had died of hypothermia.

The deadly lesson of the *Titanic*, that cold water is no place for the unprotected human body, is now so obvious that further remarks about it seem superfluous. At the time, however, the report of the investigation headed by the British shipwreck commissioner, Lord Mersey, barely mentioned hypothermia as a factor in the catastrophe (Mersey, 1912). In fact, the superintendent of the Port of Southampton recorded the cause of death as drowning for each member of the lost crew.

Not until World War II and its heavy loss of life at sea was it truly appreciated that survival in cold water requires more than simple flotation vests to prevent drowning (Keatinge, 1969). In the North Atlantic alone, where water temperature even in summer rarely exceeds 15°C, almost 30,000 Allied sailors died of immersion hypothermia between 1939 and 1945. Eyewitness accounts of ships sinking along the main Atlantic sea routes during the war suggested that men could survive indefinitely only when the water temperature was warmer than 20°C and died within six hours of immersion at 15°C and within an hour of immersion at 0°C. In 1956 Professor R. A. McCance and his colleagues at the University of Cambridge published a comprehensive analysis of the hazards to those in ships lost at sea between 1940 and 1944. They showed that 45% of people who had been aboard ships that sank in water at 5°C to 9°C died, while only 23% died when the water temperature was 20°C to 31°C. Although the analysis included all causes of death, McCance correctly attributed this rather remarkable difference to hypothermia.

## Water Temperature and Human Survival

Most of the water in the world is cold relative to the minimum temperature necessary for an immersed human body to maintain thermal balance. For the resting body this water temperature has a narrow range known as the thermoneutral zone, in which the water feels completely comfortable. The thermoneutral zone occurs at a water temperature of roughly 33°C–34°C, depending on certain physical characteristics of the individual such as subcutaneous fat thickness and body surface area. In thermoneutral water the skin vessels constrict slightly and the core temperature settles out at about 36.7°C, but metabolic rate remains constant and there is no shivering.

By generating heat from shivering or exercise, healthy people can maintain thermal balance at water temperatures well below the thermoneutral zone. The so-called critical temperature is the lowest temperature not associated with a measurable increase in metabolic rate. For most people critical water temperature is between 30°C and 33°C, although values as low as 28°C have been measured for obese people. With exercise most people can maintain thermal balance at a water temperature as low as 25°C (77°F). At temperatures below 25°C the effects of exercise, as will be discussed later, are highly unpredictable.

The warm thermoneutral zone of the human body is due to its high surface area-to-volume ratio and the high density and thermal conductivity of water relative to air. The average monthly surface water temperature of the ocean ranges from below 0°C near the poles (seawater freezes at –2.2°C, or 28°F) to about 30°C at the equator. Most of the offshore coastal waters of North America are less than 25°C for much of the year. Commercial vessels are required to carry immersion suits for operation in latitudes north of 32°N and south of 32°S in the Atlantic Ocean and 35°N or S in the Pacific Ocean. Regardless of latitude, if local water temperature is below 20°C, immersion suits should be carried at sea.

The dangers of accidental immersion in cold water are dramatically illustrated by accident statistics from the U.S. Coast Guard for workers on commercial fishing vessels. The risk of work-related death for commercial fishermen can approach 180 per 100,000 per year, and approximately 75% of the deaths are due to cold immersion when a vessel founders or a fisherman falls overboard. This death rate is more than 15 times greater than the occupational death rate for firefighters.

## Prediction of Survival Time in Cold Water

Prediction of individual survival time in cold water is an inexact science that, as some claim, is plagued by too many variables. However, assuming rescue is not immediately available, survival in cold water for a body of known size boils down to two factors: the temperature of the water in relation to the area of the exposed body and the individual's behavior in response to the immersion. In extremely cold water, when the incident is accidental and the victim has been caught unprepared, water temperature alone is a fairly reasonable predictor of survival time.

The effect of water temperature reduces to a basic physics problem, as does exposure to virtually every other extreme environment. The colder the water and the greater the body area immersed relative to mass, the faster a body cools. If body temperature reaches 29°C (85°F), the individual is incapacitated and can do nothing to help him- or herself. Other cold water problems, mentioned below, can shorten the survival time predicted by water temperature and body morphology.

All else being equal, and to state the obvious, it is always preferable to fall overboard into water above 20°C than water below 20°C. Apart from the obvious dependence of body cooling rate on temperature, other factors may hasten a person's demise in colder water, which are sometimes lumped under the term *sudden disappearance syndrome*. Sudden disappearance syndrome is encountered particularly when water temperature is below 15.5°C (60°F). The syndrome has a physiological basis in the cardiorespiratory reflex responses initiated by sudden facial immersion in cold water. These responses include gasping, profound hyperventilation, and cardiac arrhythmias. Some individuals with genetic polymorphisms of cardiac ion channels, known collectively as long QT syndrome, have a high incidence of sudden death upon immersion in cold water. In individuals with underlying cardiovascular disease, sudden immersion in cold water can be a cause of heart attack and stroke. Cold water reflex responses decrease breath holding capacity and interfere with coordination of breathing while swimming. The ability to swim in cold water is also impaired by its greater viscosity and by vasoconstriction of blood vessels in the muscles, which hastens the onset of fatigue. Thus, even superb swimmers have been known to drown within a few minutes of falling into extremely cold water.

A person who survives the initial response to sudden immersion in very cold water faces quick death from hypothermia as body heat is lost rapidly to the water. The rate of cooling for a given individual is related inversely to the temperature of the water. In other words, the colder the water, the faster the decline in body temperature. In addition, the rate of cooling is related inversely to body mass and the thickness of subcutaneous body fat: fat offers a temporary survival advantage in cold water. On the other hand, the cooling rate is directly related to the amount of exposed surface area and body parts immersed in the water. Individuals can be grouped into rapid coolers and slow coolers primarily on the basis of body fat, but in both groups the relationship between predicted survival time and water temperature follows the familiar hyperbolic curve (Fig. 11.1). For any body type physiological strain is progressively greater as water temperature falls. Most importantly, the relative position of the family of curves in Figure 11.1 indicates

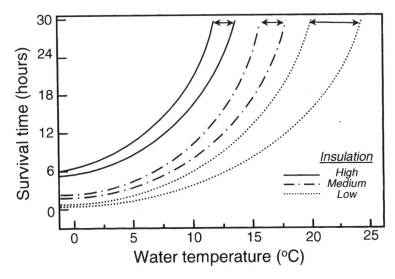

**Figure 11.1.** Human survival time in cold water. The curves indicate expected survival times at different levels of body insulation in cold water. Low insulation is light or no clothing, medium is wet suit, and high is dry suit. Dashed lines indicate ranges at low or high body mass. (Adapted from Tikuisis, Peter. 1997. Predicting survival time at sea based on observed body cooling rates. *Aviation Space Environmental Medicine* 68: 441–448.)

that the availability of a survival suit or other well-insulated clothing is far more beneficial in extending survival than is being fat (Tikuisis, 1997).

## Survival Behavior in Cold Water

Apart from water temperature, behavior in cold water is a critical independent survival factor. Why should behavior be such an important factor for survival in cold water? The short answer is the same as for exposure to most other extreme environments. If exposure exceeds the limit of a response that maintains homeostasis, it is only a matter of time before the individual dies. Thus, one faced with such a situation must behave in a way that avoids a lethal exposure for as long as possible, which requires knowledge and preparation.

Faced with immersion in cold water, preparation means being cognizant of the danger and having access to suitable protective equipment. Whether the safety equipment is adequate depends on water temperature and sea state. In some conditions, a simple personal flotation device (PFD) that keeps the head out of the water is sufficient. However, it is known from studies by the Royal Navy that cold hands and feet soon become useless for paddling, and a standard front-worn life preserver turns the face into the waves. Cold water then washes over the head and face, which increases the rate of cooling and the risk of drowning. In such conditions a full survival suit is necessary but may not suffice; sometimes only a covered raft or lifeboat and protective clothing will do.

The knowledge that body cooling in water is not uniform is a potential key to surviving a cold water immersion incident. There are body areas of particularly high and low heat loss (Hayward and Eckerson, 1984). The main physiological defense against cold water is vasoconstriction of blood vessels in the skin, subcutaneous tissue, and superficial skeletal muscles. Intense vasoconstriction provides the extremities with higher insulation values against heat loss. By comparison, the head, neck, lateral chest wall, and groin are exceptional heat loss areas. Although the superficial blood vessels of the neck, chest, and groin do constrict in response to cold, very large blood vessels, such as the carotid, femoral, and axillary arteries, course through these areas and carry so much blood from the core to the skin surface that they continue to serve as large sources of heat loss. The blood vessels of the scalp are poor constrictors, and total heat loss can increase by as much as 30% by immersion of the head. These principles are the basis for the Heat Escape Lessening Posture or HELP, in which the head is kept out of the water, the arms held next to the body, and the legs drawn up and pressed together tightly.

An important aspect of HELP behavior is to keep still. In very cold water neither shivering nor exercise, which requires movement, generate enough heat to offset the increase in heat loss that occurs by "stirring" the water around the body. The convective heat loss in moving water can be many times greater than that in still water, and there are numerous scientific reports of declining temperature with exercise in cold water at temperatures of 5°C to 15°C. However, for each body type there is a narrow range of water temperatures at which the heat generated by sustained exercise can prevent a decline in body temperature. Unfortunately, it is not possible to predict this water temperature accurately for a given person, and individual measurements are necessary to determine it. For instance, English Channel swimmers can maintain normal body temperatures for up to eighteen hours in 16°C water, whereas most swimmers require water above 20°C. The ability of the channel swimmer to stay warm may relate to the relatively thick subcutaneous layer of fat of distance swimmers who train in cold water. However, other physiological factors have not been ruled out.

Conventional wisdom about exercising to stay warm in cold water is not to do it without extra thermal protection if water temperature is 15°C or less. With exercise the rate of heat loss from the body in water can exceed 100 times that of air at the same temperature. Beyond this, it is very difficult to draw conclusions. In general, it is reasonable to expect thin people not to be able to maintain body temperature by exercising in water of 16°C to 24°C, whereas some fat people will. Also, physiological measurements indicate that body temperature is similar between rest and exercise for average-sized people in water at 25°C to 28°C and that body temperature tends to increase during exercise at water temperatures above 28°C. In any event, the best chance of surviving accidental immersion in cold water (<20°C) is to keep as much of the body as possible out of the water at all times. The only reason to swim for safety is if the distance is less than 200 yards or there is no hope of rescue.

## Hypothermia in Deep Sea Diving

The thermal problems of cold water immersion are encountered in undersea diving regardless of latitude or time of year. Worldwide, the water temperature at approximately 600 feet of seawater (FSW) is a fairly constant 4°C to 6°C. Thus, the physiological responses to body cooling and the principles of thermal protection are important factors in virtually every deep underwater diving operation. The following tragedy dramatically illustrates this problem.

In June of 1973 Clayton Link, the son of inventor Edwin Link, and an associate were exploring the ocean floor off Key West, Florida, when their submersible, the *Johnson Sea Link*, became entangled and trapped in underwater cables at a depth of nearly 600 FSW. At the time *Johnson Sea Link* was a tour de force of diving technology, a maneuverable two-man aluminum and plexiglas submarine equipped with an array of sophisticated communications equipment, scientific instruments, and safety features, including a state-of-the-art life-support system and enough oxygen for several days. For the first twenty-four hours that the two men were trapped in the aluminum bell, they were a bit cold but in good spirits. Inside the submersible $CO_2$ began to build up, and the men decided to breathe fresh air from the built-in breathing system (BIBS), which relied on a system of compressed air. Soon they became irritable and finally fell silent.

On the mother ship the support crew had mounted an immediate rescue operation that went as well as could be expected under the circumstances. The submersible was finally disentangled and hauled up after thirty-three hours underwater. When the hatch was opened they found that the cabin had been pressurized with air from the breathing system because the men had exhaled into the cabin while breathing from the BIBS air banks. Both men were dead. The loss of the two veterans was a blow to the tight-knit world of undersea exploration, and it brought into focus the irony of death from hypothermia in the subtropical waters of the Florida Keys.

Indeed, hypothermia is a major concern in the deep diving environments of military and commercial divers when extended periods are needed to complete a task. These tasks are undertaken in saturation diving, which is widely employed in the offshore oil drilling industry in the North Sea and Gulf of Mexico. Saturation divers live in a pressurized deck chamber system mounted onboard a ship or drilling platform. They work in shifts and are lowered from the deck system to the worksite in a pressurized transfer capsule, or bell. At the working depth the hatch of the capsule is opened, and the divers in their gear leave the bell to work. They are tethered to the bell by an umbilical of safety cables, communication lines, and hoses for hot water suits. The bell is run from inside by a tender who communicates with the mother ship or platform. After each shift the divers return to the bell, the hatch is closed, and the bell returns to the ship. This procedure allows shipboard personnel to tend the divers continuously and obviates the need to supply continuous power and breathing gas to great depths.

A lost bell, as illustrated by the *JohnsonSea Link* tragedy, poses a major thermal problem to deep saturation divers. *Sea Link*, however, was a true submarine; it began the day at an internal cabin pressure of one atmosphere. Thus, the gas density inside *Sea Link* was the same as was that at sea level until the men began to breathe from the pressurized breathing system and exhale into the cabin. Under these conditions the pressure increased gradually, and the men could live for no more than thirty hours.

The rate of increase and the final pressure inside the *Sea Link* are not known, but a diving bell at 600 FSW is pressurized to 19 atmospheres with helium–oxygen, so the gas inside is 19 times as dense as it would be at sea level. This dense gas also has a heat capacity 19 times as great as that at sea level. Because helium is one-seventh as dense as $N_2$ in air, survival time without supplementary heating in a pressurized bell at 600 feet, ignoring the respiratory factors noted below, would be reduced from thirty to less than eleven hours!

Heat transfer from a higher temperature to a lower temperature is required by the second law of thermodynamics in direct relation to the temperature difference between the two points. Heat transfer between the body and its environment obeys the heat balance equation, and convective losses from the skin ($C_{sk}$) account for most of the heat lost from the body during immersion in both cold water and dense gases, such as those in diving habitats. Recall that convective heat loss from the skin ($C_{sk}$) can be expressed as the convective heat transfer coefficient ($h_c$) times mean skin temperature (°C) minus ambient temperature (°C). Because the heat transfer coefficient ($h_c$) is a function of the velocity, density ($\rho$), and specific heat of the fluid (liquid or gas) around the body, it determines the tolerable ambient temperature for the body. In gas-filled underwater habitats or pressure chambers, the density of the gas is directly proportional to the pressure. As the depth increases, the convective heat loss increases directly in proportion to the temperature difference between the skin and the gas ($T_{sk} - T_a$). This heat loss problem is compensated for partly by vasoconstriction, which decreases $T_{sk}$.

The practical solution to living in dense gaseous atmospheres is to avoid the physiological responses to cold, such as shivering. This can be accomplished by increasing the temperature in the habitat as depth increases to stay in the comfort zone. The principle is shown by the list of optimum temperatures for thermal comfort in pressurized helium–oxygen in Table 11.1. At 50 atmospheres absolute (ATA) gas density is so great that the comfort zone approaches the thermoneutral zone of water.

## Respiratory Heat Losses and Slow Cooling

Respiratory heat loss hastens lethal hypothermia for underwater divers but is not encountered in cold water at sea level or inside the hull of a submarine at one atmosphere. Respiratory heat loss directly cools the body core instead of the skin. Because it does not effectively stimulate skin thermal sensors, its effects are similar to slow cooling. The dangers of slow cooling in water were brought to light primarily through research by Paul Webb in the 1970s. Webb noticed that by cooling very slowly, normal people could lose as much as 300 kilocalories without shivering. The absence of shivering negates two early defenses against hypothermia, production of extra heat and timely recognition that something is wrong; the person can cool to the point of serious fatigue or a decrement in performance or judgment without recognition of danger.

Because thermoregulatory mechanisms were not designed to detect unusual respiratory heat losses, slow cooling problems arise from breathing cold, dense gas. At 600 FSW in 4°C water, all the heat produced by metabolism is lost from the body core by ventilation alone. Furthermore, shivering begins only when brain and spinal cord temperatures begin to fall and heat loss from the lungs cannot be compensated. Failure occurs because the increase in ventilation needed to take up $O_2$ to generate heat is offset by the extra respiratory heat loss. The same situation occurs during exercise when the increase in respiratory heat loss from increased ventilation offsets new heat production. For cold, dense gases lethal respiratory heat loss can be avoided only by heating the inspired gas.

Table 11.1. Thermal Comfort Zones in Hyperbaric Helium

| DEPTH (ATA) | COMFORT ZONE (°C) |
|---|---|
| 1 | 24–26 |
| 10 | 26–28 |
| 20 | 28–30 |
| 30 | 30–31 |
| 40 | 31–32 |
| 50 | 32–33 |

ATA, atmospheres absolute.

The accidental loss of a diving bell usually cuts the power needed to heat the gas in the diver's environment. Hence, the gas in the bell rapidly cools to ambient temperature, and the trapped diver(s) is at risk of death from hypothermia. He or she will survive only as long as available passive insulation can forestall lethal body cooling from the skin and respiratory tract. Survival time is directly related to the depth (gas density), the temperature of the water, and the volume of gas needed to support pulmonary ventilation.

# 12

## Air as Good as We Deserve

The evolution of advanced forms of animal life on Earth would not have been possible without molecular oxygen ($O_2$), but too much of it is toxic to virtually all cells and organisms. Oxygen is the third-most abundant element in the universe after hydrogen and helium. It is formed at the heart of stars by the fusion of helium with carbon. More than 90% of the known universe is made up of hydrogen, whereas the other four major molecular building blocks of life, oxygen, carbon, nitrogen, and phosphorus, collectively account for less than a quarter of 1% of its composition. The abundance of hydrogen implies that the cooler spots in the universe are, in chemical terms, reducing environments. This means chemical energy is exchanged primarily by reactions that transfer electrons from hydrogen to suitable acceptors, such as carbon and nitrogen. These reductive processes are responsible for the production of many common, simple compounds in the universe, such as methane ($CH_4$) and ammonia ($NH_3$). This condition was certainly the case on Earth for billions of years, until photosynthesis appeared, which led to the generation of most of the $O_2$ present in the atmosphere today (Gilbert, 1996).

### Life in an Oxidizing Atmosphere

The earliest life forms on Earth were unicellular organisms without a nucleus, (prokaryotes), which would have been destroyed by exposure to molecular oxygen. Such organisms are strict anaerobes; they cannot survive in the presence of $O_2$, which kills them by oxidizing constitutive molecules. Nonetheless, anaerobes still thrive on Earth as descendents of these first prokaryotes, the anaerobic bacteria and the Archaea (formerly called Archebacteria). Although prokaryotes, the Archaea are not bacteria but instead share a common ancestor with bacteria (Fig. 12.1). The major phenotypes of Archaea are extreme thermophiles (heat-loving), extreme halophiles (salt-loving), sulfate reducers, and methanogens. Stratification by phenotype does not adhere strictly to genotype, in part because the methanogens have given rise to halophiles and sulfate reducers. Many thermophilic Archaea thrive in boiling water, a feature they share with thermophilic bacteria, which suggests primitive thermophilic prokaryotes as common ancestors of both. The methanogens are anaerobes that derive energy by generating methane from hydrogen and carbon dioxide in the following simple reaction:

$$H_2 + CO_2 \rightarrow CH_4 + 2H_2O$$

Today the Archaea occupy rather specialized niches because Earth's atmosphere of nearly 21% $O_2$ is highly toxic to them. In addition, even though it supplies their energy needs, too much hydrogen is also toxic. The same principle holds for $O_2$-dependent organisms; they can be killed if the $O_2$ concentration becomes too high. Consequently, Archaea reside primarily in deep-earth and deep-sea environments. Still, methanogens are responsible for most of the methane in the Earth's atmosphere today. For more than a billion years such organisms were the only life on the planet, and their evolution may be common throughout the universe.

In the history of primitive life, the Earth's biosphere made a transition from a reducing to an oxidizing atmosphere. This transition occurred gradually as early

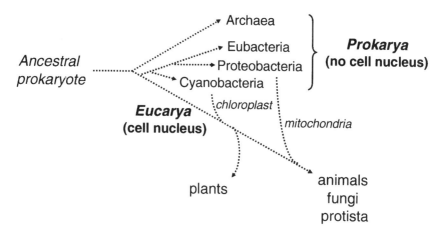

**Figure 12.1.** Evolution of eukaryotic cells containing chloroplasts and mitochondria. Chloroplasts and mitochondria arose from bacterial symbiotes after cell nuclei appeared.

photosynthetic organisms acquired the ability to use sunlight to split water, which released $O_2$ into the atmosphere and forever changed life on the planet. The earliest rocks in the geological record that contain iron oxide are approximately 2 billion years old. Although life had existed for 1.5 billion years before then, it is unlikely that a significant amount of $O_2$ was present in the atmosphere. Atmospheric $O_2$ concentration reached 10% of its present value about 1.5 billion years ago and for the last 500 million years has been stable near its current level.

Chlorophyll-based photosynthesis is found in five genera of modern bacteria, but only the cyanobacteria perform $O_2$-generating (oxygenic) photosynthesis. All cyanobacteria have this faculty, which distinguishes them from other bacteria. Thus, prehistoric cyanobacteria have long been implicated in the generation of the Earth's oxygen-rich atmosphere. Photosynthetic eukaryotes contain specialized organelles called chloroplasts or plastids that handle photosynthesis. Modern molecular genetics has indicated that chloroplasts derived from ancient cyanobacteria that were somehow incorporated into eukaryotic host cells and reverted to a simpler form, or "endosymbiont." With the loss of autonomy, chloroplasts became organelles supported by their hosts and used to harvest light energy and convert carbon dioxide and water into an organic substrate. Thus, nearly all the $O_2$ released into the atmosphere is attributable to cyanobacteria, from both plants and algae and the bacteria themselves.

Most eukaryotes are able to respire with $O_2$ because they contain mitochondria. Mitochondria consume more than 90% of the $O_2$ of animal cells to produce energy. As with chloroplasts, molecular sequencing indicates that mitochondria are derived from an ancient bacterial endosymbiont of Proteobacteria. Some eukaryotes lack mitochondria because they either were never acquired or were lost along the way. In either case, eukaryotes without mitochondria are anaerobic organisms.

The abundance of $O_2$ in the atmosphere has had several important effects on living organisms and is directly responsible for the evolution of vertebrate animals. The presence of $O_2$ allows energy to be stored in the biosphere, and the molecule is thermodynamically poised to participate in many chemical oxidation–reduction reactions. When electrons are transferred to $O_2$ and it is reduced to water, energy is liberated. By accepting electrons $O_2$ acts as an oxidizing agent, thus altering the structure of the donor molecule, including enzymes and other constitutive biomolecules. In turn, biomolecules have evolved to take advantage of the energy source, and new enzymes that contain active iron or copper have evolved to accelerate the relatively sluggish reduction of oxygen. Thus, the life cycle of the biosphere settled into its present mode: photosynthetic splitting of water into $O_2$ and $H_2$ for energy, followed by the opposite process, respiration, in which hydrogen from hydrocarbons is used to reduce $O_2$ to water and liberate energy. This simple and elegant cycle, driven by sunlight, is the source of energy for most of the life on the planet.

It is important to recognize that the cycle of the biosphere is not closed but contains leaks that lead to losses of $O_2$ and $H_2$ from the system. Fortunately, these losses are replenished by natural planetary sources, which for the foreseeable future will more than offset the losses, but there remains the inexorable loss of hydrogen to interstellar space and the possibility of gradual increases or decreases in the concentration of $O_2$ in the atmosphere. Throughout geological time atmospheric $O_2$ concentration has fluctuated, reaching as much as 30% during at least one era in which the photosynthetic biomass was unusually high.

## Biological Oxidations and Oxygen Toxicity

The abundance of $O_2$ in the atmosphere increases the rate of oxidation of many molecules that donate electrons to it. Oxidations in living systems, or biological oxidations, are important products of normal biochemical processes and toxic byproducts of others. Oxidation occurs when molecular $O_2$ accepts electrons from any of a variety of donor molecules. These reactions oxidize the donor and reduce $O_2$ to either a reactive oxygen species (ROS) or to water. The complete reduction of $O_2$ to water requires four electrons, so incomplete reduction with one, two, or three electrons leaves $O_2$ in a reactive state. The rate of incomplete $O_2$ reduction increases as its concentration increases in the environment.

The most important medical problem of excessive $O_2$ exposure is oxygen toxicity (Halliwell amd Gutteridge, 1999). Its existence was first proposed by the chemist Joseph Priestly, who codiscovered oxygen in 1775 at about the same time as did Scheele. Priestly called oxygen "dephlogisticated air" because he believed a substance he called "phlogiston," which we know was nitrogen, had been removed to make a purer form of air. In a remarkably prescient statement, Priestly noted, "Although pure dephlogisticated air might be very useful as a medicine, it might not be so proper for us in the usual healthy state of the body; for as a candle burns out much faster in dephlogisticated than common air, so we might live out too fast and the animal powers be too soon exhausted...."

More than a century later, in 1878, the French scientist Paul Bert, using a pressure chamber, discovered $O_2$ toxicity of the central nervous system (Bert effect). In 1899 J. Lorrain-Smith discovered pulmonary $O_2$ toxicity, and in 1944 $O_2$ damage to the retina of the newborn was discovered, a condition known as retrolental fibroplasia. All three expressions of $O_2$ toxicity can be significant clinical problems. For the brain and the lungs, the relationship between $O_2$ dose and duration of exposure follows the familiar rectangular hyperbola that describes the approach to many other survival limits (Fig. 12.2). The asymptote of the curve indicates the maximum safe dose for the human lung at sea level is between 40% and 60% oxygen. The curve asymptote for the human brain is well into the range of hyperbaric pressures, close to 1.5 ATA. It is important to recognize that $O_2$ toxicity is related to the absolute partial pressure of oxygen ($O_2$), not to the percentage of oxygen in the atmosphere.

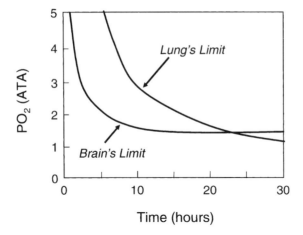

**Figure 12.2.** Oxygen toxicity limits for the brain and lungs. The effect curve for exposure time against dose follows a hyperbolic relationship, which is unique for different organs and tissues depending primarily on the extent of antioxidant defense and repair mechanisms.

The first evidence that $O_2$ toxicity is mediated by free radicals was published in 1954, when Rebecca Gerschman and colleagues noted that the pattern of X-radiation damage to lung tissue was similar to that induced by high $O_2$ concentrations. The essential feature of this free radical theory of damage was that $O_2$ toxicity and radiation injury involve a common mechanism mediated by the generation of radicals. In 1969 Irwin Fridovich and Joseph McCord demonstrated the potential threat of $O_2$ radicals in biology when they discovered an enzyme, superoxide dismutase (SOD), that scavenged the free radical superoxide anion, thereby eliminating it from further reactions. Their work provided direct evidence that $O_2$ radicals were produced in vivo.

A free radical is defined as any chemical species that contains at least one unpaired electron. During normal respiration mitochondria completely reduce $O_2$ to water by the addition of four electrons and two protons. When the $O_2$ molecule is reduced incompletely, for example, with one, two, or three electrons, ROS are generated, including superoxide ($O_2^-$), hydrogen peroxide, and the hydroxyl radical (·OH), the most powerful oxidant in biology. The chemical formation of ROS is illustrated in Figure 12.3. ROS are generated at many sites in eukaryotic cells, which in some cases use them for metabolic activities or as cell signals. ROS are produced from 1% or 2% of the $O_2$ consumed by normal cells. For instance, certain plasma membrane-associated enzymes, such as the NADPH oxidases*, produce $O_2^-$

---

*NADPH oxidases are enzymes that use electrons from a donor, nicotinamide adenine dinucleotide phosphate (NADPH), to convert $O_2$ to superoxide anion ($O_2^-$).

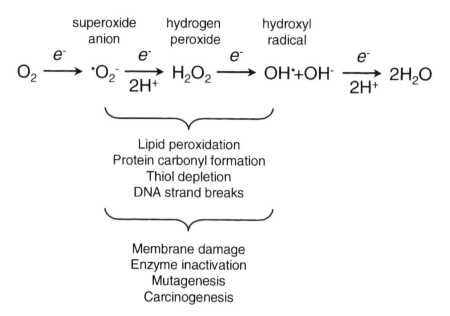

**Figure 12.3.** Production of reactive oxygen species (ROS) by incomplete reduction of molecular oxygen ($O_2$). Normally, only 1% to 2% of the $O_2$ used by the body generates ROS, but in stress, disease, or injury the production rate increases. When production of ROS exceeds the capacity of the antioxidant defense mechanisms, damage occurs to biological macromolecules such as proteins, lipids, and DNA.

as a precursor of secondary oxidants, hydrogen peroxide and hypochlorous acid, which are used to kill microbes by white blood cells.

In the 1980s another free radical, nitric oxide (NO), was found to be involved in the normal dilation of blood vessels. First known as endothelial-derived relaxing factor, or EDRF, NO is made in the lining of the blood vessel, the endothelium, where it diffuses into the smooth muscle layer of the of blood vessel wall and binds to the enzyme guanylate cyclase. NO stimulates guanylate cyclase to produce the messenger molecule cyclic guanosine monophosphate (GMP), which relaxes smooth muscle and lowers blood pressure. In contrast, lack of NO rasies blood pressure, and this appears to be involved in certain forms of hypertension. It was soon found that NO is formed in many mammalian cells from the amino acid L-arginine via the actions of a family of $O_2$-enzymes called nitric oxide synthases. NO has also turned out to be an important mediator of many other physiological processes, including neurotransmission, platelet aggregation, cell survival, and immune defenses, as well as the flashes of fireflies.

NO is a rather stable radical, with a biological lifetime on the order of seconds. It reacts with a limited number of macromolecules, in particular proteins that contain reduced sulfur (thiols) and heme moieties such as those at the active sites of hemoglobin and cytochromes. Excessive NO production is toxic due to the formation of secondary reactive species derived from its reactions or those of its chemical products with ROS. NO and its chemical products are known as reactive nitrogen species (RNS). Thus, NO plays a complex role in biology because it is involved in both normal and disease processes.

## Antioxidant Defenses and the Oxidant–Antioxidant Balance

To combat $O_2$ toxicity, mammalian cells maintain abundant defenses. These include special enzymes, radical scavengers, and mechanisms to repair oxidative damage. The major antioxidant enzymes include three types of SOD and two peroxide-metabolizing enzymes, catalase and glutathione peroxidase. The major scavengers include water-soluble compounds such as ascorbic acid (vitamin C), uric acid, reduced thiols such as glutathione, and lipid-soluble compounds such as flavonoids, carotenoids (such as vitamin A), and tocopherols (vitamin E). These defenses are maintained in specific locations in cells where unregulated biological oxidations are undesirable.

When the production of ROS exceeds the ability of antioxidant defenses to contain them, nonspecific oxidation reactions occur that are deleterious to the cell. This disrupts the normal oxidant–antioxidant balance and is known as oxidative stress. Unchecked oxidative stress alters biological processes by modifying the structure and function of proteins, lipids, and nucleic acids. Augmentation of antioxidant defenses, either through natural increases in gene expression or by supplementation, protects against oxidative stress, including $O_2$ toxicity and radiation injury. Oxidative stress is also associated with the damaging effects of air pollution, diseases such as arteriosclerosis and Alzheimer's, and natural processes such as aging.

For many years oxidative stress was considered entirely toxic and antioxidants purely protective. It is now recognized that oxidants contribute to homeostasis by regulating biochemical events such as gene expression, neurotransmission, and cell growth. The discovery of specific genes and cell signaling reactions affected by oxidants has led to the belief that they are important biological signals. Additionally, antioxidants modulate the activity of numerous genes and cell signaling events. Multiple, diverse inflammatory and toxic events and their effects on cells are mediated either directly or indirectly by these pathways, which contain elements sensitive to chemical modification by oxidants.

The study of untoward biological oxidations began in a rather unusual way. Oxidation of fat, known as lipid peroxidation, was first studied in order to understand why butter and other dietary fats so quickly become rancid. Eventually, this work yielded important insights into the role of oxidized lipids in cell biology. The major targets of lipid peroxidation in the cell are plasma membranes and lipoproteins. Lipid peroxidation is frequently initiated when free radicals remove or abstract hydrogen from carbon atoms of polyunsaturated fatty acids. In the presence of $O_2$, this process can propagate as a chemical chain reaction. Lipophilic antioxidants such as vitamin E and the carotenoids are able to terminate these lipid chain reactions.

Lipoprotein oxidation has been implicated in the etiology of arteriosclerosis. In lesions of arteriosclerosis cholesterol accumulates due to the uptake of oxidized low-density lipoprotein (LDL) by a scavenger receptor (SRA). Oxidized LDL is toxic to vascular lining (endothelial) cells by inducing superoxide generation and impairs NO-dependent vessel relaxation. It also stimulates the growth and proliferation of several cell lines, including tissue macrophages and smooth muscle cells, and stimulates the attraction of damaging inflammatory cells to the vessel.

Oxidative damage to DNA is repaired by a variety of mechanisms that exist in all aerobic cells. However, when too much oxidative DNA damage accumulates it causes mutation and carcinogenesis. Cancer induced by cigarette smoking, certain chemicals, and chronic inflammation is related to oxidative DNA damage. Oxidants form chemical adducts with DNA bases and sugars and cause DNA strand breaks. Then DNA cross-links with proteins and other macromolecules. If not repaired, these defects cause permanent misreading of the genetic information in the cell's descendants.

## The Free Radical Theory of Aging

What is the relevance of this oxygen chemistry to human survival? For years it has been known that animals with high metabolic rates have shorter life spans and age more rapidly than do animals with slow metabolic rates. It also has been observed that caloric restriction slows aging and prolongs life in animals. These findings have been interpreted to support a free radical theory of aging, first proposed by Denham Harman in the 1950s. Harman postulated that energy consumption was related to senescence and referred to the idea as the "rate-of-living" hypothesis (Harman, 1956). He suggested that free radicals produced as a by-product of respiration caused cumulative oxidative damage that results in aging. Harman noted parallels between aging and ionizing radiation, including mutations and cancer. It is also relevant to note the relationship between metabolic rate and survival of vertebrate animals placed in pure $O_2$ environments (Fig. 12.4). In general, cold-blooded, long-lived animals such as turtles are resistant to $O_2$ toxicity, while short-lived animals such as rodents die rapidly.

The realization that ROS are produced in mitochondria during respiration linked the rate-of-living hypothesis to the damaging effects of free radicals. It was proposed that a higher rate of respiration leads to faster generation of ROS and promotes aging. Indeed, superoxide production by mitochondria is sufficiently great

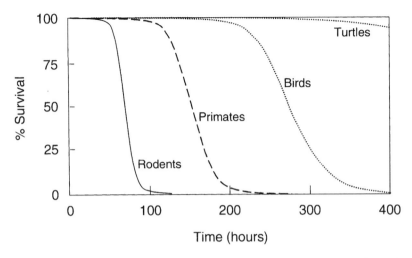

**Figure 12.4.** Survival curves for vertebrates exposed to pure oxygen ($O_2$). Death from hyperoxia is due to a combination of lung injury and other factors. The survival function varies in rough proportion to the specific metabolic rate of the body.

to require protection by a unique SOD, MnSOD, which converts it to $H_2O_2$. MnSOD is expressed normally as well as following many types of stress, including infections.

The relationship between oxidative stress and aging has been reinforced by measurements of oxidative damage to DNA and protein that indicate ubiquitous damage to proteins and DNA and a trend for damage to increase with age (Sohal et al., 2002). Indeed, the amount of oxidized protein and DNA inside mitochondria can increase almost exponentially with aging, and the persistence of these products appears to be associated with leakage of $H_2O_2$ from the organelles into other parts of the cell.

Among the changes of senescence in many species, including humans, is the gradual accumulation inside cells of a yellow–brown pigment called lipofuscin, an insoluble residue of cross-linked lipid and protein created by the oxidation of cell membranes. Other cross-linkages are also formed between proteins, and the presence of cross-linking is a marker of aging. Another type of oxidative cross-linkage is the formation of bonds between tyrosine amino acid residues in proteins, or bi-tyrosine generation, originally described in the lignification of woody plants. Other markers of protein oxidation, such as protein carbonyl groups (C=O), also increase in mammalian cells with aging. For example, carbonyl content increases from 10% of the total protein pool in the young to 30% in older subjects. This correlation has been interpreted as evidence for the role of oxidative stress in the aging process (Castro and Freeman, 2001).

As time passes it is not the direct effects of age that cause trouble but the development of certain diseases. The most important causes of death in an aging population are a few diseases that have a basis in oxidative stress. The incidence of death from these diseases increases with age according to the familiar hyperbolic relationship (Fig. 12.5). This suggests that aging is associated with the failure of normal homeostatic processes for cardiovascular health and immune function. It appears likely, but unproven, that oxidative stress plays an important role in the loss of these homeostatic mechanisms.

Premature senescence is characteristic of some rare genetic diseases, such as progeria and Werner's syndrome. These diseases have provided important clues to aging processes. Children born with these syndromes age a lifetime in a decade or less. In children with Werner's syndrome, the tissue level of protein carbonyls is similar to that of octogenarians. This aging-related oxidant accumulation supports the idea that senescence is associated with ROS generation and a decline in antioxidant capacity. For some reason the repair mechanisms designed to prevent accumulation of the detrimental products of oxidative damage are no longer sufficient.

Conditional support for the rate-of-living hypothesis also comes from the association of calorie restriction with longevity in animals. Indeed, calorie restriction is the only intervention known to delay aging in mammals. Many such studies, although difficult to conduct, have confirmed that low calorie intake without malnutrition significantly increases the life span of rodents (Merry, 2000). The reasons for the effect and its origins, however, have been elusive. An interesting evolutionary hypothesis is that this represents adaptation to famine, which shifts resources away from reproduction and toward homeostasis (see Shanley and

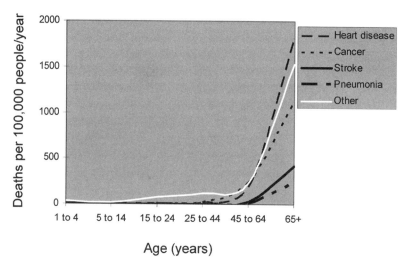

**Figure 12.5.** Approximate mortality rates as a function of age in the U.S. population. Death rates for cancer and heart disease increase exponentially with age, which is reflected in all-cause mortality rates. (Data excerpted from 1998 U.S. government mortality statistics.)

Kirkwood, 2000). The animal gains an increased survival probability by reducing its rate of senescence and preserving its reproductive capacity until the famine ends.

Whether calorie restriction has similar effects in longer-lived species more closely related to humans remains controversial. Calorie restriction and aging studies in rhesus monkeys have been consistent with the rodent studies. Monkeys on calorie restriction have less body fat, lower body temperature, lower blood glucose levels, lower insulin levels, and lower lipid levels. In addition, the response to insulin infusion is increased in these monkeys. Studies of caloric restriction at the National Institute of Aging also have provided some evidence that the strategy may reduce risk factors for certain age-related conditions such as diabetes and cardiovascular disease.

The reasons why caloric restriction retards aging are unknown, but all calorie restricted animals do not show decreases in metabolic rate. However, calorie-restricted animals do appear to be resistant to oxidative stress. A link between oxidative stress and calorie restriction has been proposed involving blood sugar and insulin resistance. According to this hypothesis, which is still incomplete, hypersecretion of insulin because of impaired glucose metabolism promotes aging by stimulating mitochondrial oxidant production. Insulin resistance is found in obesity, adult onset diabetes, and about a third of adults after a carbohydrate load. Insulin may also promote cellular aging by inhibiting degradation of oxidized protein and by stimulating polyunsaturated fatty acid synthesis. Mutations in the insulin receptor that lead to severe insulin resistance in rodents result in shortened lifespans. However, mice lacking the insulin receptor only on fat cells have increased longevity despite normal calorie intake, suggesting that leanness and not food restriction is the more important factor in lifespan (Bluher et al., 2003).

It is worth noting, that in kwashiorkor, ironically, the dire appearance of famine edema is strongly associated with oxidative stress. This oxidative stress is due to the depletion of the antioxidant glutathione and the accumulation of reactive iron and lipids in mitochondria. However, after refeeding kwashiorkor survivors show no evidence of either premature aging or unusual longevity. This apparent inconsistency highlights our limited understanding of the oxidative link to aging, beyond the simple fact that metabolism and oxidative stress are integrally woven into the biochemical mechanisms at multiple levels.

# 13

## Bends and Rapture of the Deep

In the early history of diving, the lack of underwater technology limited descent to the depth that could be reached with a single breath of air. As diving technology appeared and improved, the limit was pushed downward until new physiological limits were encountered. As interest grew in deeper diving, biomedical research overcame apparent limits, but new limits appeared, until ultimately the effects of hydrostatic pressure per se seemed to stop humans from descending farther. Thus, the history of diving is a paradigm for the relationships among technology, physiology, and human factors in the effort to conquer extreme environments (Phillips, 1998). This chapter describes how two important physiological limits have been dealt with in diving from scientific and engineering perspectives: decompression sickness and the effects of pressure on the nervous system. Before discussing these problems it may be helpful to review the natural limits of breath-hold diving.

Breath-hold diving was practiced for centuries primarily to collect sponges and pearls and recover valuable objects lost in shallow water. Even today, practiced divers routinely descend to sixty feet and hold their breath for two minutes. The record for human breath-hold diving, with descent rate artificially assisted by weights, is in excess of 500 feet of seawater. However, this assisted maximum breath-hold depth and the maximum breath-hold time are modest compared to diving mammals.

All diving mammals hold their breath, and their breath-hold time is limited by the amount of $O_2$ available in the body at the start of the dive. In diving mammals

$O_2$ is stored bound to myoglobin in muscle, but in humans, tissue storage of $O_2$ is minimal and the lungs hold virtually all the $O_2$ needed for the dive. Some diving mammals also remain submerged substantially longer than useable $O_2$ should be available. They extend their dives by gliding to save energy, by selective body cooling, and by redistributing blood flow and $O_2$ supply away from nonessential organs to the heart and brain. The latter response is the so-called diving reflex, which, although present in humans, is not well developed.

During a breath-hold, as $O_2$ is consumed carbon dioxide ($CO_2$) is produced and enters the blood. Because there is no ventilation, the partial pressure (P) of $CO_2$ gradually rises. The length of the breath-hold is determined by the time it takes for blood $PCO_2$ to increase to a critical point, known as the breakpoint. Within a minute or two, the action of $CO_2$ on the respiratory centers in the brain forces the diver to breathe.

As a breath-hold diver descends water pressure squeezes the thorax, and gases in the lungs, primarily $O_2$ and $N_2$, are absorbed in increased amounts into the blood. As the diver ascends the chest re-expands, and the gas in the lungs also expands. This dilutes any $O_2$ remaining in the lungs as the diver approaches the surface, and blood $PO_2$ falls more rapidly than normal. Thus, the breath-hold diver is in danger of losing consciousness from hypoxia at the end of the ascent. This event, known as shallow water blackout, illustrates why it is dangerous to hyperventilate to extend breath-hold time. Hyperventilation decreases the initial blood $PCO_2$, thus providing more time to reach the breakpoint, but it also allows more time for $PO_2$ to fall to dangerously low levels (Stolp et al., 1997).

Breath-hold diving is an inefficient and dangerous way to work underwater, and attempts to develop diving apparatus date back thousands of years. It is generally agreed that the first effective diving systems were diving bells, the initial practical embodiment of which was developed in 1717 by Royal Astronomer Edmund Halley. Effective compressed air diving apparatus, however, required the invention of the steam-powered air compressor in the late eighteenth century.

## Decompression Sickness

In the mid-nineteenth century, it was discovered that breathing compressed air at pressures greater than about twenty feet of seawater can be associated with decompression symptoms, originally known as caisson disease or bends. Caisson disease and bends refer specifically to the joint pain that afflicted workers in dry underwater tunnels or bridge encasements after they ascended from a work shift. Bends is the most common and notorious form of decompression sickness, although more disabling decompression illnesses such as spinal paralysis and arterial gas embolism can occur during too-rapid ascents (Edmonds et al., 2002).

Decompression sickness is initiated by the formation of bubbles of inert gas, such as nitrogen, in body tissues and blood after ascending from a higher to a lower pressure. The risk of decompression sickness increases as a function of the amount of inert gas taken up while at depth and eliminated from body tissues during the ascent. This inert gas uptake increases as the depth and duration of the diving exposure increase. Thus, safe decompression requires adherence to a decompression schedule, or table, which often includes safety stops as the diver ascends toward the surface. The longer and deeper the exposure, the greater the decompression obligation incurred.

Decompression sickness is a direct consequence of the physical behavior of gases in aqueous solution because the human body is mostly water. As a diver descends in the water column, hydrostatic pressure increases by one atmosphere absolute (ATA) for each 33 feet of seawater (FSW). Thus, at 33 FSW the total pressure is 2 ATA. In order to inhale against this water column, the pressure of the diver's breathing gas must be increased in proportion to the pressure. As a result, more gas molecules occupy the lungs and other gas-containing cavities of the body at constant volume and temperature. The relationships of pressure (P), volume (V), and temperature (T) to the number of moles of gas (n) is described by the ideal gas law:

$$PV = nRT$$

where R is the universal gas constant. The special gas laws most relevant to diving, such as Boyle's law ($P_1V_1 = P_2V_2$) and Charles's law ($P_1/T_1 = P_2/T_2$), are derived from the ideal gas law.

Air and other breathing gases are mixtures of $O_2$ and other gas molecules. In a mixture of gases, the total pressure is the sum of the partial pressures of each of the gases (Dalton's law). This means each component of the mixture behaves as though it alone occupies the available space. The uptake of gas by body tissues is determined primarily by the diffusion of gas into or out of blood from the lungs. The amount of a gas dissolved in liquid at any temperature, such as blood or tissues of the body at 37°C (98.6°F), is also proportional to its partial pressure (Henry's law).

Because air is 78% nitrogen ($N_2$) and 1% argon (Ar), most of the gas breathed into the lungs is biologically inert. This inert gas is dissolved in blood plasma in proportion to its partial pressure in the lungs and its solubility in blood in accordance with Henry's law. The dissolved gas is carried in the circulation to the tissues, where it is taken up according to the same physical principles. The rate of inert gas uptake is determined primarily by blood flow (perfusion) to the tissues and by the gas's solubility in the tissue. Thus, $N_2$, which has high fat (lipid) solubility, is taken up more quickly by the brain, which has a high lipid content and high blood flow. Each "tissue" can be thought of as a compartment having halftimes for the uptake and elimination of gases. Tissues with high rates of blood flow tend to have short halftimes, and those with low rates of blood flow have longer halftimes. For example, a tissue with a halftime of five minutes will take up gas six times as fast as will a tissue with a thirty-minute halftime.

During decompression the gas-exchange process is reversed, and gas re-enters the circulation and is eliminated from the body through the lungs. The rate of elimination of inert gas from the body therefore represents a composite of all the tissue halftimes. A model of inert gas elimination using multiple exponential halftimes has served admirably as the basis for decompression tables since the work of J. S. Haldane nearly 100 years ago. If the external pressure drops too rapidly, however, gas in the tissue comes out of physical solution and forms bubbles. This situation has been likened to opening a bottle of soda and releasing the $CO_2$ gas from solution (hence "soda pop"). In the body, however, the gas bubbles are composed primarily of $N_2$, and the bubbles that form can obstruct the circulation and initiate biochemical injury to cells and tissues. Symptoms arise most often from the joint spaces and the fatty tissues of the nervous system; the former have slow gas release rates and the latter have high gas uptake rates.

Decompression sickness is also associated with rapid ascent to altitude, such as flying in unpressurized aircraft. The altitude bends problem is essentially the same as that of divers. The liquid in an aviator's body contains an appropriate amount of $N_2$ for sea level, but with rapid ascent to 18,000 feet (or above), where the barometric pressure is half that at sea level (0.5 ATA), this $N_2$ can leave solution and form bubbles in tissues. This problem was encountered in U.S. Army aircrews at the start of World War II when all Allied aircraft were unpressurized and many routinely flew above 25,000 feet. At these altitudes supplemental $O_2$ had to be provided to the crews to prevent performance decrements from hypoxia. However, $O_2$ not only prevented the hypoxic effects of altitude, it accelerated $N_2$ elimination and decreased the risk of decompression sickness.

In the United States plans for a pressurized high-altitude bomber designed to evade antiaircraft fire were on the drawing board as early as 1933. The prototype evolved into the Boeing B-29 Superfortress, but the aircraft did not enter service until September 1942, in part because it had a service ceiling of more than 30,000 feet, and a major concern of flying pressurized aircraft in combat, the effects of explosive decompression on the human body, were unknown. Ultimately, however, some 2000 of these bombers were employed in the Pacific theater. Two, the *Enola Gay* and *Bock's Car*, dropped atomic bombs on Hiroshima and Nagasaki in August 1945 (Figure 13.1).

The problem of explosive decompression was encountered when a projectile penetrated the cabin and decompressed it in a matter of seconds. This exposed the crew to the dangers of decompression sickness, gas embolism, and hypoxia. The problem was addressed at the beginning of the war in a series of medical experiments at Wright Field in Dayton, Ohio. One difficulty quickly surpassed all others in importance: useful time after explosive decompression was limited by unconsciousness, which occurred within a minute or two, depending on the altitude (Fig. 13.2). Thus, the limiting factor for surviving explosive decompression turned out not to be decompression sickness, which, although common, painful and potentially disabling, was rarely fatal. Rather, rapid loss of consciousness was due to the very low $O_2$ tension in the air at such high altitudes, which produced cerebral hypoxia. The problem could be avoided by having a mask supply oxygen for use on the plane much like on today's commercial airliners. The use of supplemental $O_2$ for high altitude flights is covered in Chapter 17.

In underwater diving with compressed gas, decompression sickness tends to be more severe than it is on ascent to altitude because the amount of inert gas available to leave solution after the dive is much greater. The partial pressure of the $N_2$ in tissues may be many atmospheres, depending on the depth and duration of the dive. As deadly as this can be, safe decompression practice has eliminated decompression sickness as the factor that limits human survival underwater.

The ability to dive deeper with exotic breathing gases, such as mixtures of helium and oxygen, led to longer decompression obligations relative to working time

**Figure 13.1.** The B-29 Superfortress was the first aircraft to fly in combat with a pressurized cabin. The bomber's service ceiling of more than 30,000 feet enabled it to fly into the stratosphere, above antiaircraft fire. During World War II some 2000 of these bombers were employed in the Pacific, and two, *Enola Gay* and *Bock's Car*, dropped atomic bombs on Hiroshima and Nagasaki in August 1945. (Copyright The Boeing Company, reproduced with permission.)

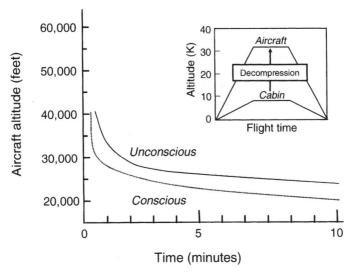

**Figure 13.2.** Time of consciousness after explosive decompression of pressurized aircraft at high altitude. Curves are predictions based on individual data points from exposures of a small number of nonacclimatized healthy subjects at different altitudes. The gray area indicates the range of individual variability and loss of mental acuity without loss of consciousness. The inset shows the protective effect of cabin pressurization on cabin altitude during a typical commercial airline flight.

for divers. To solve this problem, in 1957 U.S. Navy medical officers led by George Bond explored a new approach to extend the time underwater. After twelve to twenty-four hours under pressure, the tissues of the body become saturated with the inert gases in the breathing mixture, which means they are unable to take up any more inert gas unless the pressure is increased further. Thus, the obligation for decompression reaches a plateau: a diver at that depth can work for long intervals and then return to the surface using a single, slow decompression schedule that is independent of the duration of the exposure. This is known as saturation diving. Indeed, it is even possible for saturation divers to make excursions up and down in the water column, within certain predefined limits, without risk of decompression sickness. Using the principles of saturation, divers have lived and worked for months at ocean depths of 1500 FSW and at simulated depths of 2400 FSW in dry chambers.

## Rapture of the Deep

There are physiological effects of inert gases on the body that very clearly limit the ability of humans to live at great depth. As underwater pressure increases to 4

ATA or more, compressed air divers begin to experience euphoria or feelings of intoxication. The euphoria and the impairment in judgment and intellectual performance accompanying it are known as rapture of the deep or nitrogen narcosis. The manifestations of rapture of the deep are directly related to the partial pressure of $N_2$ in the body. Hence, its manifestations increase in direct proportion to ambient pressure. This problem is not encountered by breath-hold divers, who, as noted above, can descend to remarkable depths, albeit for very short periods of time.

In a single breath of air drawn at sea level, the total $N_2$ contained in the lungs, no matter how deep a diver goes, is insufficient to cause either decompression sickness or nitrogen narcosis. However, narcosis does appear rather quickly when descending on compressed air because a substantial partial pressure of $N_2$ is achieved in the brain. $N_2$ uptake produces euphoria similar to that produced by alcohol and by the early stages of general anesthesia. For the diver narcosis clouds the mind, predisposes to poor decisions, and causes disorientation to time, depth, location, and direction. The depth of onset and intensity of narcosis is commonly described by martini's law: each fifty feet of depth is approximately equal to drinking one martini on an empty stomach. At air pressures beyond 12 ATA (363 FSW), narcosis may be severe enough to cause unconsciousness, thus placing a physiological limit on the useful depth of compressed air as a breathing gas. Many record-seeking scuba divers have tested their mettle against nitrogen at more than 400 feet and failed with deadly consequences, a practice akin to playing Russian roulette with an increasing number of bullets in the revolver.

Narcosis is a biological effect of $N_2$ caused by all noble gases, such as neon, argon, and xenon, and by gaseous anesthetics. Noble gases occupy the same column on the periodic table of elements because their outer shells are filled completely with electrons. This makes them chemically inert, which means they are taken up and excreted unchanged. The narcotic properties of these inert gases correlate well with their solubility in body fats or lipids (Table 13.1). For example, helium, which is the least lipid-soluble noble gas, has a narcotic potency one-fourth that of nitrogen, while xenon is highly lipid-soluble and about twenty-five times more narcotic than $N_2$. Carbon dioxide, although not a noble gas, also has signifi-

Table 13.1. Relative Narcotic Potencies of Inert Gases

| INERT GAS | MOLECULAR WEIGHT | OIL/WATER SOLUBILITY | NARCOTIC POTENCY |
|---|---|---|---|
| Hydrogen | 2 | 2.1 | .56 |
| Helium | 4 | 1.7 | .24 |
| Neon | 20 | 2.1 | 0.28 |
| Nitrogen | 28 | 5.2 | 1.0 |
| Argon | 40 | 5.3 | 2.3 |
| Krypton | 84 | 9.6 | 7 |
| Xenon | 131 | 20 | 25 |

cant narcotic properties and at one time was erroneously proposed as the cause of diving narcosis.

Despite considerable research, the precise biological mechanisms of inert gas narcosis are not well understood. Many scientists have favored mechanisms that involve the physical state of the nerve cell membrane, the so-called critical volume hypothesis, which has long been used to explain effects of volatile general anesthetics (Bennett, 1993). The essence of the hypothesis is as follows. Molecules of anesthetic gas are absorbed into the lipids of the membranes of the nerve cells (neurons), thereby causing them to increase slightly in volume. This increase in volume alters the electrical and ionic permeability characteristics of the membranes, particularly the cell projections, or axons, which branch out to adjacent neurons. Volatile gases with very high lipid solubility, such as nitrous oxide ($N_2O$), achieve the critical volume effect at lower partial pressures than do gases such as $N_2$. This hypothesis is often quoted because lipid-soluble gases are better general anesthetics. The critical volume hypothesis, however, is, at best, only a part of the narcosis mechanism.

General anesthetics act by altering the electrical activity in certain regions of the brain and spinal cord. Several anatomical brain structures have been suggested as the primary sites of anesthetic action. In the brain stem a group of cells (predominantly neurons) called the reticular formation plays a key role in the state of consciousness and in regulating motor activity. Suppression of the reticular formation has been implicated in the effects of general anesthesia, but anesthetics also interfere with neuronal activity at other sites, including the cerebral cortex and the hippocampus, which is involved in memory. Anesthetics also affect the pathways that connect different brain regions, such as those that handle sensory traffic from the thalamus to the brain cortex.

At the level of individual neurons, anesthetics usually depress the excitability of axons, which transmit electrical signals along the membranes by ion movement in the process of neurotransmission. These membrane effects may alter the behavior of nerve synapses, the sites of contact between adjacent neurons. By influencing axonal membranes, anesthetic and narcotic gases may affect presynaptic release or postsynaptic response to neurotransmitter molecules in the brain, thereby interfering with cell-to-cell communication.

Recent evidence suggests that narcotic and anesthetic gases actually interact primarily with membrane proteins, perhaps directly at the synapses themselves. For instance, certain anesthetic gases, such as halothane, prolong the effects of inhibitory neurotransmitter amino acids in the brain, such as gamma-amino butyric acid, or GABA, which opposes the influence of excitatory amino acids. Inhaled anesthetics that alter synaptic function typically have a much greater effect on synapses than on impulse conduction by axons. In any event, it is not yet known how the lipid solubility of a gas and its protein effects interact at specific locations to produce a state of anesthesia (or narcosis) in the brain.

A good deal of research has been done to attempt to determine whether tolerance or adaptation to inert gas narcosis occurs in humans. These studies have shown that the decrement in cognitive performance on breathing inert gases under pressure varies greatly among individuals, but repeated exposures of the same person to the same environment generally produce roughly the same decrement in performance. Small improvements in performance are generally attributable to learning. In other words, scientific evidence is lacking that humans adapt significantly to the effects of inert gas narcosis. The same is true of general anesthetics. Unless an anesthetic gas can be metabolized or chemically altered by the body, there appears to be no mechanism by which the body can adapt to it. This makes intuitive sense from an evolutionary perspective, because exposures to high partial pressures of inert gas are purely the product of human technology, and there are no known natural selective forces in the environment that would favor human adaptation or cross-acclimation to narcosis or anesthetic gases.

## Pressure Reversal of Anesthesia and the High-Pressure Nervous Syndrome

Any hypothesis that explains the mechanism of narcosis also should explain the ability of very high pressures to reverse the effects of general anesthesia. Take, for instance, the following famous experiment. When an anesthetic gas such as nitrous oxide ($N_2O$) is pumped into the water around a tadpole, the animal soon falls motionless to the bottom of the tank. If the pressure inside the tank is increased hydraulically to 100 times normal, the tadpole promptly revives despite the presence of the anesthetic gas. In other words, increasing the hydrostatic pressure or compressing the tadpole with a gas of low narcotic potency, such as helium, abolishes the effects of certain general anesthetics and the narcotic properties of inert gases under pressure. This phenomenon is known as pressure reversal of anesthesia. Although not all animal species show pressure reversal of anesthesia, in small mammals such as the mouse, a hydrostatic pressure of 100 ATA will increase the anesthetic dose requirement by roughly 50%. Pressure reversal of anesthesia appears to be related to a general phenomenon of increased excitability of the central nervous system produced by the effects of high pressure. It is not known whether pressure reversal represents antagonism at specific sites of anesthetic action or generally counteracts the depression of neuronal activity by anesthetics.

In practical terms, the effects of high pressure on the nervous system have posed a barrier to rapid compression and very deep diving (Bennett and Rostain, 2003). Rapid compression and high pressure have untoward effects known collectively as high-pressure nervous syndrome (HPNS), although these two aspects of the problem may have different mechanisms. Humans and other primates generally begin to experience HPNS when the pressure is greater than 20 ATA (666 FSW).

The physiological basis of the syndrome is poorly understood, but it may involve decreased fluidity of cell membranes, altered neurotransmitter release, or postsynaptic effects of pressure.

The hallmarks of HPNS are rapid tremor, poor coordination, involuntary jerking movements, and microsleep. These manifestations correlate with abnormalities in an electroencephalogram. The brain's alpha wave activity slows, and its theta wave activity increases. The severity of HPNS increases as a function of the pressure to which the body is exposed, and at depths beyond 1000 FSW people often suffer from debilitating tremors, nausea and loss of appetite, shortness of breath, and disconcerting nightmares. In animals, depending on the species and individual susceptibility, extremely high pressures cause convulsions and death. Even the nervous systems of creatures such as squid, adapted to life at great ocean depths, show the effects of HPNS if the pressure is increased to high enough levels. Although there are individual differences in susceptibility to HPNS, as with nitrogen narcosis, there is no scientific evidence that physiological adaptation to pressure occurs in human beings. The molecular events appear to be mediated by physical processes to which human beings have had no reason to adapt.

The addition of a small amount of a narcotic gas in a breathing mixture decreases the effects of HPNS, which, according to the critical volume hypothesis, would allow cell membranes to assume a more normal conformation. This hypothesis assumes that nitrogen narcosis and HPNS operate by exactly or almost exactly opposite effects. This notion is probably too simplistic, but the approach has resulted in the successful use of trimix (helium, nitrogen, and oxygen) for very deep human dives. For example, trimix containing 5% nitrogen in the Atlantis series under the direction of Dr. Peter Bennett at Duke University in the early 1980's allowed men to live and work successfully at a simulated depth of 2250 FSW. In the 1990's divers at Comex in Marseille reached a simulated depth of 2400 FSW using a breathing mixture of four gases: helium, nitrogen, hydrogen, and oxygen. At the present time, however, HPNS, more than any other single physiological factor, limits the maximum depth to which humans can dive.

If the absolute limit for human survival under high pressure is scaled in proportion to that of lower primates, irreversible convulsions in humans should occur at approximately 3330 FSW. This pressure, approximately 100 times that at sea level, probably represents the limit of tolerance of the human nervous system. Although some individuals in the population are more resistant to the effects of pressure and may be able to surpass this limit, there are no obvious forces of natural selection that would be expected to assist acclimation of humans to such extraordinary atmospheric pressures.

A few words should also be said about the density of the breathing gas as a potential limit to human survival at extreme pressures. As depth increases the density of the gas increases in proportion to the pressure. Thus, at 100 ATA breathing gas density is 100 times that at sea level. Dense breathing gas increases the resistance to respiratory gas flow, both in airways and in external breathing circuits. Ultimately, as pressure increases gas density becomes too high to permit the respiratory muscles to move the gas and ventilate the lungs. This density limitation appears first with exercise, when high ventilation is needed to support metabolism and is a serious practical impediment to liquid breathing. However, the problem is circumvented to some degree by nonnarcotic gases such as helium and hydrogen, which are much less dense than air. These breathing gases mixed with $O_2$ will support adequate ventilation, at least for light work, at depths of perhaps as much as 5000 FSW.

How do humans compare to other species with respect to tolerance to high pressure? Clearly, many invertebrates and fishes are capable of living at extraordinary depths in the ocean, but their nervous systems are far simpler than are those of humans. However, some highly intelligent diving mammals are capable of withstanding pressure more than twice as great as can humans without experiencing HPNS. For example, sperm whales descend without a hint of HPNS to 6000 FSW and elephant seals to 5000 FSW with an ease and alacrity that is truly astonishing. The mechanisms by which these animals have adapted to hydrostatic pressure are essentially unknown, but, based on the ability of drugs that alter synaptic neurotransmission to ameliorate HPNS in some species, it is reasonable to speculate that nerve synapses have been a site of pressure adaptation. Animals with such enhanced diving capabilities have been favored most likely because it provides such an important survival advantage with respect to foraging for food.

## Implications of High Pressure for Human Life on Other Planets

The implications of high-pressure biology for adapting to life on other planets are clear; not only are composition and temperature critical, but the pressure of the atmosphere humans can tolerate is tightly constrained. If an extraterrestrial atmosphere contained large amounts of narcotic gases such nitrogen or argon, atmospheric pressure would have to be within a factor of roughly ten of that of Earth. If the atmosphere on a hypothetical planet contained large amounts of a nonnarcotic gas such as helium or hydrogen, people could expect to survive exposure to pressure approaching 100 times that of Earth. Beyond these rough limits the human central nervous system functions rather poorly.

An interesting thought experiment is to contemplate the possibility of life under the atmosphere of Earth's "hothouse" sister planet, Venus. Venus has a surface atmospheric pressure ninety-six times that of Earth, which is close to the theoretical limit for human tolerance. To put this into perspective, a spaceship on the surface of the planet would need the strength of the hull of a modern nuclear submarine to avoid being crushed by the dense atmosphere of Venus. The atmosphere also contains 97% carbon dioxide ($CO_2$), no $O_2$, and very little water. This enormous concentration of $CO_2$ has produced a runaway greenhouse effect that makes Venus as hot as an oven (surface temperature >450°C).

To transform Venus into a habitable place would first require the colossal undertaking of absorbing virtually all the atmospheric $CO_2$ in order to reduce the planet's temperature and atmospheric pressure to levels compatible with life on Earth. For mammals, including humans, breathing $CO_2$ at even a single atmosphere of pressure is so narcotic that deep anesthesia occurs in a matter of minutes. In fact, people can tolerate $CO_2$ breathing only at a pressure equivalent to roughly one-tenth of an Earth atmosphere. In other words, even if the atmosphere of Venus contained enough $O_2$ to support life, the amount of $CO_2$ is close to 1000 times the maximum tolerable concentration for the human body. The pressure, temperature, and composition of the atmosphere of Venus may be fairly typical of Earth-sized planets that orbit close to G-type stars. If so, the challenges of finding other habitable planets in this part of the Milky Way will be formidable.

# 14

## Sunken Submarines

The Barents Sea is one of the most inhospitable maritime environments in the world. Along the northern coastline of Scandinavia, the tumultuous surface freezes solid for four months every winter. In the peak of summer surface water temperature is 5°C, cold enough that a lightly clad man in a life vest would survive less than an hour. Nonetheless, the Barents Sea provides the only major sea route for commercial shipping into and out of western Russia during the warmer months. It is also home to the Russian Northern Fleet, whose warships sail from Murmansk with relative lack of scrutiny compared to more temperate ports on the Baltic and Caspian Seas.

Sailing with the Northern Fleet, despite modern advances in maritime technology, is almost as perilous today as it was a century ago. Since World War II the peacetime Soviet Navy has lost dozens of ships and many hundreds of sailors in the sea, including four nuclear submarines. The most recent tragedy occurred on August 12, 2000, when a state-of-the-art nuclear submarine sank with all hands lost, sending the Russian people into a state of national mourning.

### The Sinking of the *Kursk*

While on a tactical naval exercise in the Barents Sea, the *Kursk* (K-141), an Oscar-II Class submarine carrying 118 officers and men, was shattered by two

**Figure 14.1.** Oscar-II Class Russian submarines, one of eleven, the *Kursk* was lost in the Barents Sea on August 12, 2000, with 118 men aboard. The *Kursk*'s sister ship, the *Omsk*, is shown here at sea. Initially, twenty-three men survived a pair of explosions in the forward torpedo room that drove the submarine into the bottom in 456 feet (137 meters) of water. Unfortunately, the survivors could not be rescued.

massive underwater explosions, the second of which reached 3.5 on the Richter scale and sent the submarine crashing into the seabed in 456 FSW (137 meters) at 69°40" N, 37°35" E off the Kola Peninsula. The *Kursk* was one of eleven Oscar-II Class submarines, the largest nuclear attack submarines in the world and the most sophisticated vessels ever operated by the Soviet Navy (Figure 14.1). These enormous undersea boats represent masterpieces of Soviet Cold War technology, capable of carrying up to twenty-four supersonic cruise missiles, twenty-four torpedoes, and a dozen subsurface mines. Commissioned in 1994, the *Kursk*'s double-hull construction and nine isolatable compartments were believed to make it virtually unsinkable.

The *Kursk* had a submerged displacement of 24,000 tons and measured approximately 155 meters (513 feet) long by 18 meters (60 feet) wide and had a draft of nearly 10 meters (33 feet). The boat could dive 2000 feet, reaching speeds in excess of 30 knots, and had sufficient power to remain underwater until the rations for the crew ran out. Powered by twin 190-megawatt nuclear reactors, the *Kursk* could generate enough electricity to power 100,000 conventional homes.

The submarine was named for the eleventh-century city on the famous Kursk salient of the central steppes, the site of the greatest tank battle of World War II.

The battle of Kursk has achieved near mythical standing in the eyes of the Russian people. It began on July 4, 1943, with the Red Army facing the mightiest German force ever assembled, consisting of 3000 tanks, 1800 aircraft, and nearly a million men. The Nazi high command had planned the offensive for nearly four months, biding their time until they had amassed enough men and material to ensure a decisive victory. Meanwhile, however, the Red Army had laid out unprecedented defenses in preparation for the battle. They matched the Germans in men and held numerical superiority in artillery and armor.

In battle the German panzers were repelled by 4000 Soviet tanks, with the loss of some 50,000 German and 200,000 Soviet lives. Wherever the German army advanced, it met ferocious resistance, and on July 12 in a climactic encounter near the village of Prokhorovka, the Russians destroyed more than 300 German tanks. On July 13, after nine days of desperate but inconclusive fighting, the battle was halted by Adolf Hitler himself. The Germans retreated, ending the last major offensive on the Eastern front and the last hope for a Nazi victory over Russia.

The loss of a submarine with the hallowed name of *Kursk* fixed the collective Russian mind on the disaster and on theories about its cause. The investigation showed that the initial explosion in the submarine had punched a two-by-three-meter hole through the hull between the first two compartments of the boat and caused a fire in the forward torpedo room. Two minutes and fifteen seconds after the first blast, a second explosion occurred equivalent to several tons of TNT, and a minute and forty-five seconds later the *Kursk* plunged bow-first into the bottom, flooding the forward three-quarters of the submarine.

In the final analysis it has been agreed that the second explosion was related to the detonation of munitions in the forward torpedo room, possibly conventionally armed *Shkval* class torpedoes. The submarine sank so fast that no distress signal was sent, and the crew never had the opportunity to float its emergency communications beacon. The Russian Navy initially claimed a collision by an unidentified foreign submarine caused the first explosion, but in inner circles it was known that just beforehand, the captain of the *Kursk*, G. P. Liachin, asked to jettison a malfunctioning torpedo. It was later acknowledged that the first explosion was due to detonation of the torpedo propellant and the second to most or all of the complement of warheads.

Mercifully, most of the men of the *Kursk* were killed instantly. This much is known from salvage of the submarine, which revealed that the massive second explosion had shattered the first three compartments beyond recognition. It was also learned that twenty-three men survived the accident in the two aft compartments of the stricken submarine for an indeterminate period of time.

On October 26, 2000, while recovering the bodies of the submariners inside the *Kursk*, a handwritten note was found in the pocket of lieutenant Dimitri Kolesnikov. According to Russian officials, Kolesnikov's note was written between 13:34 and 15:45 on August 12, six to eight hours after the explosions on the day of the accident. The note revealed that most of the men from compartments VI, VII, and VIII had moved to compartment IX in the rear at 12:58 and were considering trying to make an ascent, two or three at a time, from the rear escape hatch. Russian officials have never released additional details, but unofficial reports indicate that another entry may have been made on paper some time on August 15, three days later. Later, another note was found on another dead sailor, which confirmed Kolesnikov's note.

The final lucid moments of the 23 survivors, as they pondered the situation in their dark, steel tomb, must have been filled with a bewildering collage of conflicting emotions. One can imagine the disbelief and shock, terror and self-control, anger and serenity, and, in the end, the hopelessness they faced. For most people their plight is unthinkable; for submariners and their families it is unspeakable. The men of the *Kursk* faced the most difficult decision that can ever be made by the crew of a submarine: attempt to escape or await rescue.

## The Debate over Submarine Escape

Sunken submarines are a stark reality of the silent service. Since World War II submarines arguably have been the most important warships on the high seas. They patrol not only major shipping lanes but coastlines and harbors of opportunity. It is in these shallow waters, where most losses have occurred, that the best opportunity for rescue exists. The issue of whether to escape or await rescue on a disabled submarine has been debated hotly for decades by the international submarine community.

In principle, submarine escape in shallow water, from depths of 100 FSW or less, is relatively straightforward. In the U.S. Navy thousands of twentieth-century American submariners were trained to escape from submarines in water-filled towers in New London and Honolulu. The Navy submarine escape program was abandoned in the 1980s because it was believed to expose trainees to an unnecessary risk of cerebral air embolism, the deadly release of gas into the circulation from overpressurization of the lungs during rapid ascent. In addition, rescue programs were viewed as more practical. The British Royal Navy, however, continued to develop and refine submarine escape procedures, and today the self-rescue philosophy has been revived in the U.S. Navy. In theory, escapes from U.S. and British submarines are now possible from depths of at least 600 FSW, and Royal Navy submarine escape suits have been tested successfully to this depth (Brown, 1999).

In the *Kursk,* on the seabed under 456 FSW, escape was theoretically possible but practically unrealistic because Russian submarines are not equipped with proper escape suits or properly designed escape hatches. Furthermore, the water was too cold to make an escape attempt without survival suits. The men who survived the explosions on the *Kursk* were forced to hunker down and await rescue from the surface.

How did these twenty-three men, in some ways ultimately less fortunate than their ninety-five shipmates, die? How much did they know about their fate or their chances for survival? Is it reasonable to think they could have been rescued? What factors, in general, define the limits of human survival in a disabled submarine? It is fascinating, in a macabre sense, that Kolesnikov's entry on his notepaper, made in the dark at 15:45 on the day of the accident, estimated their chances of survival at 10% to 20%. It will become apparent that his estimate is testimony to the phrase "Hope springs eternal."

## The Physics of Submarine Disasters

Modern nuclear submarines are designed to operate at tremendous depths, the limits of which are determined by the strength of the inner hull. The inner hull, also known as the pressure hull, protects the environment of the crew, which lives and operates inside the boat at a normal barometric pressure of one atmosphere (ATA). Outside this steel hull the column of seawater exerts an extra atmosphere of pressure (14.7 pounds per square inch, psi) for every 33 FSW. Thus, at a depth of 1000 feet, the pressure difference across the inner hull exceeds 31 ATA, or 460 psi. The inner hulls of submarines of the U.S. Navy today have test depths of more than twice this value.

What happens inside a submarine when the integrity of the inner hull is breached? The consequences of these terrible scenarios are easy to envision: if the breach is structurally catastrophic, such as when the submarine sinks below its crush depth, the hull implodes, the air escapes, and everyone is killed in a matter of seconds. This was precisely the situation with the loss of two U.S. nuclear submarines in the Atlantic in the 1960s. Although the causes of these catastrophes were different, both vessels exceeded crush depth and imploded. SSN *Thresher* sank in 1963 in almost 8000 FSW and *Scorpion* in 1968 in nearly 13,000 FSW. At such extreme pressure there is no hope for the crew. The hull crumples and disintegrates as it spirals to the bottom. In general, no provisions for survivors are made either by crew or by rescue personnel. These accidents are simply investigated, and, if possible, the boats are salvaged with the hope of identifying a failure that can be prevented in the future.

In times of peace the loss of submarines in such deep water is the exception rather than the rule. Most submarines sink in shallow water due, in large measure, to collisions with other vessels. When the breach in the hull is not structurally catastrophic, water gradually floods the internal compartments and compresses the air inside the submarine until cabin pressure is equal to water pressure outside. At this point flooding stops, and the volume occupied by water is proportional to the depth. In some cases flooding can be controlled before it critically compromises the cabin atmosphere, but the boat's electrical and propulsion systems may become disabled by the accident. This describes the case of the most famous submarine rescue in history, the recovery of thirty-three sailors from the U.S. Navy submarine *Squalus*, which sank in 1939 off the coast of New England in 243 FSW.

The men of the *Squalus* were brought to the surface in a rescue chamber known as the McCann bell, which was attached with a cable to the escape hatch by divers. Under the direction of Swede Momson, commanding officer of the U.S. Navy Experimental Diving Unit, the McCann bell made four trips to the disabled submarine to bring up survivors. At the time mounting a rescue attempt in 243 FSW pushed the limits of diving technology.

The water where the *Kursk* came to rest off the Kola Peninsula was more than 200 feet deeper than that in which the *Squalus* went down. The 456 FSW above the *Kursk* exerted a pressure on the hull equivalent to approximately 15 ATA. Normally, this depth would have been the no problem for *Kursk*, which, like other Russian submarines, routinely operated below 400 feet. However, at that depth a two-by-three-meter hole in the pressure hull allowed seawater to enter the submarine so rapidly that it sank in a matter of seconds. As the submarine hit bottom, unless the watertight doors between compartments had been sealed, water would have continued to pour into the boat until all but one-fifteenth of the cabin was flooded, and the air in the nonflooded space would be compressed to 15 ATA.

Details of the disaster suggest the air pressure inside the *Kursk* was increased but never reached 15 ATA. The bodies of the twenty-three original survivors were found in the two aft compartments of the vessel. The notes written by Kolesnikov and his companion confirmed the high atmospheric pressure inside the cabin, but the exact cabin pressure that was reached is unknown. Assuming the internal volume of the two aft compartments was proportional to the average size of the seven flooded compartments, less than 80% of the vessel was flooded. However, in order for the pressure to rise to 15 ATA, 93.3% of the cabin volume would have to have been flooded.

Using these assumptions and Boyle's law, the pressure inside the rear two compartments can be calculated as

$$P_1 \times V_1 = P_2 \times V_2$$
$$15 \times 0.067 = P_2 \times 0.20$$
$$P_2 = 5.03 \text{ ATA}$$

Thus, the air pressure in the aft compartments of the *Kursk* after the disaster was probably about 5 ATA. It may have been even less if the watertight doors in the rear were closed before the forward part of the hull was completely flooded. In addition, had the survivors been breathing air at 15 ATA, they would have suffered dire physiological consequences almost immediately. They likely would have been unconscious within minutes and dead within hours. Incapacitation would have been caused by nitrogen narcosis and death by $O_2$ toxicity.

The conclusions of this simple analysis of the *Kursk* disaster are derived as follows. If the pressure inside the *Kursk* had risen to 15 ATA, the amount of nitrogen in the air would have been 15 × 0.79, or 11.85 ATA. The remaining 3.15 ATA would have been primarily oxygen, so within minutes the survivors would have been incapacitated by nitrogen narcosis and shortly thereafter would have, died of oxygen toxicity. Certainly, no one would have been able to write a lucid note in the cold blackness of a stricken vessel after several hours at 15 ATA. On the other hand, the amount of nitrogen at 5 ATA is 5 × 0.79, or 3.95 ATA, and $O_2$ is 1.05 ATA, assuming it had not been consumed in the explosion and fire. The narcotic effect of air at 5 ATA is tolerable, and the survival time for healthy people breathing oxygen at 1 ATA is approximately seven days.

Breathing air at 5 ATA can be tolerated long enough to mount a rescue attempt. In fact, the U.S. Navy submarine rescue program is intended to bring up a submarine crew from just such a depth. Unfortunately, the Russian Navy has no such capability, and in the first four days after the sinking the Russians managed to get only a few photographs of the *Kursk's* hatch. Nevertheless, it is important to determine whether the twenty-three men of the *Kursk* could have been rescued if proper equipment had been available. Such an analysis helps determine if the criticism of the lack of Russian diving and rescue capability was justified in the aftermath of the accident.

## Analysis of Survival Factors on Sunken Submarines

Most submarine engineers use a formal approach to the problem of survival aboard sunken submarines based on differential equations that predict the rate of change of critical parameters in the cabin atmosphere. A simpler approach based on steady state values is sufficient to understand the nature of the problem. The loss of the *Kursk*, although no one was rescued, provides an opportunity to understand the practical problems of submarine rescue.

To assess the chances for rescue of the twenty-three survivors of the *Kursk*, an approximate survival time must be determined for the crew. As noted above, the first step is to compute an actual volume and atmospheric pressure inside the nonflooded compartments and then account for the critical atmosphere parameters inside the sunken submarine. The critical parameters are:

1. cabin volume
2. cabin temperature
3. atmosphere pressure and composition

4. toxic gases, such as carbon monoxide (CO) or chlorine, in the atmosphere
5. amount of $CO_2$ generated by metabolism
6. amount of $O_2$ required by the survivors
7. atmosphere control equipment and supplies available to the survivors

These factors are considered individually in the following paragraphs. We will assume no radiation exposure and that the concentration of CO or other toxic gases from the explosion was not very high. If true, the time of consciousness on the *Kursk* was limited primarily by the rate of metabolic $CO_2$ production, and death was from $CO_2$ narcosis and hypothermia.

First, consider why the supplies of water and food aboard the sunken submarine were not critical factors. Chapter 5 indicated that potable water is more important for survival than is food. As a general rule, survival time without water is four to seven days, far less than the eight weeks it takes to starve to death. Aboard a modern submarine, freshwater stored in tanks is accessible from almost anywhere on the boat. The presence of a single full forty-gallon hot water heater would roughly double the survival time for twenty-three men with respect to water.

Second, it is important to consider the note written by Kolesnikov's companion that indicated the survivors were poisoned with CO in the explosion and fire. Partial asphyxiation from CO, other toxic gases, or other chemical reactions is almost certain. Furthermore, the huge explosion would have consumed a lot of $O_2$ and could have decreased the cabin supply enough to shorten survival time. That one or both of these factors were critical determinants of survival, however, is doubtful for the following reasons.

Russian submarines are equipped with efficient absorption systems for toxic gases that do not require electrical power, that is, they are either passive or operated by hand. On U.S. submarines this measure is sufficient to deal with most atmosphere control emergencies, although any massive cabin explosion would pose a serious challenge. U.S. Navy submarines carry portable atmosphere monitoring devices, both active and passive chemical absorption systems, and built-in breathing systems, or BIBS, to provide clean air at all times. BIBS are operated from high-pressure air banks located throughout the submarine. These requirements are part of a rigorous, mandatory atmosphere quality program aboard every American submarine. This is also true of Russian submarines, but without more information and an analysis of the composition of the final atmosphere on the *Kursk*, the possibility of death from explosion gases cannot be excluded. The poisoning explanation has been put forward publicly by Russian naval and political authorities, but salvage information indicates some survivors may have lived as long as three days. This would be unlikely if they had been poisoned fatally with fumes.

A third step in this grim analysis is to estimate the actual volume of the two compartments that held the twenty-three survivors. Because the blueprints for

Oscar-II Class submarines are classified, it is difficult to say with confidence just how these compartments were configured, and some assumptions must be made. In a 500-foot submarine with an internal diameter of 30 feet, a reasonable estimate of cabin gas volume, accounting for the space occupied by machinery and equipment, is half an actual cylinder volume. This is $500 \times \pi r^2$, or $\frac{500}{2 \times 3.14 \times 225}$, just less than 177,000 cubic feet. If the two compartments that were not flooded were roughly 20% of the original volume, the amount of air trapped in them would be $177,000 \times 0.20$, or 35,400 cubic feet. This volume of gas was probably available to the crew. The estimate of cabin pressure of 5.0 ATA after flooding is consistent with the entire original volume of air in the boat being compressed into two compartments, that is, $35,400 \times 5.0$ ATA, or 177,000 cubic feet.

If 177,000 cubic feet of air was available to twenty-three survivors, the $O_2$ supply would have been at least $177,000 \times 0.21$, or ~37,200 cubic feet. This would have been sufficient for a long time because each man would be expected to consume no more than 0.5 liter of $O_2$ per minute, or slightly more than 1 cubic foot per hour. Thus, 24 cubic feet of $O_2$ per hour, or 576 cubic feet per day, would be used to support twenty-three survivors, and a sixty-four-day supply was trapped in the cabin. However, consciousness would not be possible if the $O_2$ percentage fell much below 10% of actual cabin volume. This means 3540 cubic feet of $O_2$ was not useable, which cuts just over 6 days off the survival time. Therefore, a conservative estimate of the $O_2$ supply aboard the *Kursk* is 57 days.

What about the possibility of $O_2$ toxicity at 5 ATA? The survivors would have started out breathing the equivalent of 100% $O_2$ at sea level, which is clearly toxic to the lungs. This $O_2$ concentration is not lethal for about a week and would have fallen 2% per day due to consumption by the crew. Thus, the men of the *Kursk* could have survived in this $O_2$-rich environment for one week, but it then probably would have become too stressful (equivalent to 86% at sea level) for anyone to have adapted and survived until the supply was exhausted.

Because pulmonary $O_2$ toxicity would not have caused death for a week, a more stringent limit must be sought. The large amount of $O_2$ at 5 ATA also would have opposed asphyxia by CO because $O_2$ competes with CO for binding the hemoglobin molecule. This suggests that $CO_2$ build-up or hypothermia would have been more immediate threats to survival. Indeed, the main reason ventilation is so critical in submarines is to eliminate the $CO_2$ produced by the crew.

The $CO_2$ produced by metabolism is roughly equal to the $O_2$ consumed. The ratio of metabolic $CO_2$ produced to $O_2$ used is known as the respiratory quotient, or RQ. RQ varies from approximately 0.7 to 1.0 depending on the foodstuff (substrate) being used, or oxidized. When protein is oxidized RQ is 0.7, and for pure carbohydrates, such as glucose, it is 1.0. The latter number means that a liter of $CO_2$ will be produced for each liter of $O_2$ consumed by the body.

To compute survival time on a sunken submarine, a worst-case analysis is always made. Thus, the highest possible value is used for the RQ (1.0) to compute the $CO_2$ production rate of the crew. At RQ of 1.0 the amount of $CO_2$ produced per day by the twenty-three men would be 24 cubic feet per hour, or 576 cubic feet per day. At this point one must determine how far $CO_2$ concentration can rise in a cabin before causing incapacitation and death.

U.S. Navy submarines maintain a constant cabin $CO_2$ concentration below 1% throughout patrols. This limit was set years ago by research at the Navy Submarine Medical Research Laboratory in New London based largely on the efforts of one man, Dr. Karl E. Schaefer. Schaefer was a German scientist who worked on U-boat life support and cabin atmosphere control during World War II. After the war he came to the United States and worked as a civilian scientist for the U.S. Navy for the rest of his career. He turned his attention to life support in nuclear submarines and made numerous measurements of $CO_2$ concentration and its physiological effects on submariners on long patrols (Schaefer, 1958). Schaefer was among the first to recognize the importance of controlling $CO_2$ in submarine atmospheres. His 1% limit for $CO_2$ is approximately one-tenth the concentration that causes unconsciousness in humans.

Aboard the *Kursk*, the $CO_2$ concentration before the accident is unknown, but the $CO_2$ limit in Russian submarines is similar to that of American submarines. Assuming the $CO_2$ concentration in the cabin of the *Kursk* was 1% at the time of the accident, it could have increased tenfold before it became a lethal problem for the crew. At a volume of 35,400 cubic feet, the aft compartments of the submarine would have been able to handle approximately 3540 cubic feet of $CO_2$ before reaching 10% by volume. At a production rate of 576 cubic feet per day, the concentration of $CO_2$ in the atmosphere would have reached 10% in only six days.

The analysis has now encountered a new limitation to survival on the *Kursk*. The $CO_2$ concentration in the atmosphere could have reached life-threatening levels in less than a week unless the submarine were equipped with chemical $CO_2$ removal capability. In U.S. submarines this problem is handled with lithium hydroxide, which has a high capacity to absorb $CO_2$ and can be used to scrub air directly or with hand held pumps. Theoretically, each gram of lithium hydroxide absorbs 0.919 grams of $CO_2$, although in practice the efficiency is about 50%. This amounts to roughly 215 liters of $CO_2$ per kilogram of lithium hydroxide. Thus, to absorb 576 cubic feet of $CO_2$ (~16,245 liters) and extend the survival time for twenty-three men by one day, 75.5 kilograms (166 pounds) of lithium hydroxide crystals would have been needed. Even if more than a one-day supply had been available for twenty-three men in the rear of the submarine, chemical $CO_2$ absorption by lithium hydroxide requires heat, and, as will be seen, the submarine would have been cold.

To estimate the temperature of the air inside the cabin of the disabled submarine, the temperature of the water must be known. Using water temperature and air density, an estimate can then be made of how much the cabin atmosphere facilitated body cooling, just as for a survival analysis in cold water. Finally, as the men cooled, they would have begun to shiver, increasing the rates of $O_2$ uptake and metabolic $CO_2$ production. Because $CO_2$ is a more stringent limit than $O_2$, shivering shortens survival time by decreasing the time for $CO_2$ concentration to reach criticality.

Seawater temperature at 456 FSW is cold throughout the oceans due to poor penetration of sunlight to that depth. In the Barents Sea this temperature is 2°C to 4°C. Thus, the air in the sunken submarine would have gradually cooled over several hours to just above this temperature, depending on the amount of residual heat given off by the reactors. Air at 5 ATA has a heat capacity five times that of normal, which would have increased the cooling rate fivefold. The crew would have added extra clothing to stay warm, but, as demonstrated by the *Sea Link* disaster, bodies would have continued to cool from heat loss by the lungs. As the men shivered metabolic $CO_2$ production would have risen, increasing atmospheric $CO_2$ levels and shortening the time of consciousness in proportion to the extent of shivering. Normal shivering under such conditions could easily have doubled $CO_2$ production and more than offset any gain from onboard $CO_2$ absorbers. Furthermore, the temperature inside the submarine would have been too cold for effective absorption of $CO_2$ by lithium hydroxide. Thus, the cold would have shortened survival time by two or three days, and incapacitation by $CO_2$, followed by hypothermia would have caused death.

The maximum survival time aboard the *Kursk* appears to have been approximately four days and perhaps less, depending on how intensely the men shivered. Certainly, rescuers could have hoped for no more than four days for the helpless men on the floor of the Barents Sea. What resources does it take to mount a bona fide rescue effort in 456 FSW in less than four days? Is it realistic for a crew trapped at that depth to expect help, or should they attempt to rescue themselves?

Submariners have long debated the answer to these questions. Since 1937, when Swede Momsen listened helplessly for three days as the desperate tapping from the sunken U.S. submarine *S4* faded away, *time* has been recognized as the essence of submarine rescue. Sadly, no one was saved from the ill-fated *S4*, but the tragedy spurred Momsen to develop a legitimate submarine rescue program for the U.S. Navy, which moved forward on two fronts. Momsen invented a breathing device for free ascent from depth and, with McCann, a rescue chamber that could be attached to a submarine hatch by a diver, which and was used in the dramatic rescue of thirty-three men from the *Squalus*.

Swede Momsen knew that time was critical, and he understood the only real alternatives for submarine rescue. A rescue system must be poised to fly immediately to the scene of an accident, or the crew must have the equipment and capacity aboard to rescue themselves. Tragically, the Russian Navy had neither capability. A proven fly-away rescue system and highly trained and experienced personnel must be prepared to respond to any emergency, day or night, 365 days a year. Such rescue systems are valuable only if they can be deployed rapidly anywhere in the world and operate at any depth and location on the continental shelf. Such a vigilant submarine rescue capability is an expensive undertaking. By the time the Russian commanders recognized their lack of preparedness and called for help, it was too late to save the men of the *Kursk*. Although the U.S. Navy prides itself on its rescue capability, the American system has yet to be tested on an actual disabled submarine.

The alternative for the men of the *Kursk* would have been to attempt escape, but in the year 2000 this would have meant outfitting Russian submarines with the complete British escape system, including escape trunk, gas supply, and suits capable of supporting egress from up to 500 FSW. Only then provided they had been trained properly and the accident had not damaged the escape trunk, could the survivors have rescued themselves. The Royal Navy suit is dry and well insulated and offers good thermal protection while one is waiting to be plucked from the icy seas. It must be noted, however, that increased air pressure inside the submarine would have posed a risk of serious decompression sickness had the men attempted such an escape.

# 15

# Climbing Higher

Most of the world's population lives at or near sea level, where their ancestors lived for thousands of generations. Because the human body is adapted to life at sea level, it must make physiological adjustments to the decrease in atmospheric pressure at higher altitudes. These adjustments are true physiological adaptations that appear to have evolved out of the survival advantage that tolerance to $O_2$ deprivation affords the body. The amount of $O_2$ in the atmosphere declines as altitude increases, exposing the body to hypoxia, which produces the same effects as certain disorders of cardiopulmonary function.

## The Physical Environment of High Altitude

At sea level the air column above Earth exerts a force approximately equivalent to the weight of a column of mercury (Hg) 760 millimeters (29.9 inches) high. This height of mercury, placed in a barometer, exactly counterbalances the normal sea level pressure of Earth, 1 ATA, or 1 bar (1000 millibars). On ascent to altitude barometric pressure falls because the atmosphere is less dense owing to the lower weight of the air column above it. Atmospheric pressure falls more rapidly at attitude than might be predicted from the weight of the air column because as one ascends there is less compression of the air from the gas above it. Thus, atmospheric pressure falls almost exponentially with altitude (Fig. 15.1), but the composition of air remains remarkably constant between sea level and an altitude of 300,000 feet (90 kilometers, 56 miles).

At an altitude of 18,000 feet (5,486 meters) barometric pressure falls to half the sea level value, and at the summit of Mt. Everest it is about one-third the sea level value. Because the $O_2$ percentage (20.9%) in the atmosphere remains constant with altitude, the number of $O_2$ molecules in a given volume of air decreases in proportion to the decrease in barometric pressure. The partial pressure of oxygen ($PO_2$) is determined by Dalton's law, which states that the total pressure exerted by a mixture of gases in a constant volume is equal to the sum of the partial pressures of each of the components in the mixture. Thus, as altitude increases the $PO_2$ in air falls in proportion to the barometric pressure.

At sea level the pressure is 760 millimeters Hg, and the $PO_2$ in dry air is 20.9% of the total, or approximately 159 millimeters Hg. Atop Everest, where barometric pressure averages approximately 253 millimeters Hg, the $PO_2$ of the air is only 53 millimeters Hg. Interestingly, this atmospheric pressure is very close to the limit of human tolerance in terms of support of metabolic requirements. Not until 1978 did Reinhard Messner and Peter Habeler climb Everest without supplemental oxygen. Indeed, the atmospheric pressure on Everest can be tolerated only because the peak is located at latitude 28°N, which provides the climber the advantage of the equatorial bulge in atmospheric pressure caused by the accumulation of dense cold air in the stratosphere above the equator. This effect increases the

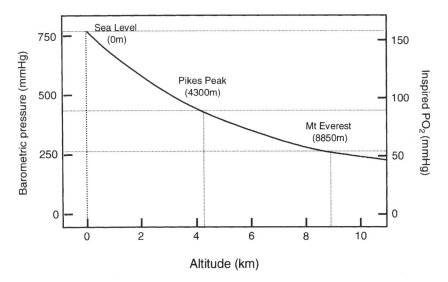

**Figure 15.1.** Atmospheric pressure and $PO_2$ fall exponentially during ascent to altitude. The fall in $PO_2$ is caused by the fall in pressure because air composition is independent of altitude, and $O_2$ remains constant at 20.9%.

atmospheric pressure on the summit by about 7% compared to the prediction of a standard atmosphere. Furthermore, small seasonal and temperature-related changes in barometric pressure can be advantageous or disadvantageous to the Everest climber.

## Physiological Responses to High Altitude

A reduction of $O_2$ in the air causes changes in many physiological processes in virtually every organ system in the body. These changes are initiated by cellular hypoxia and are generally of two kinds: rapid accommodations mediated by constitutive $O_2$ sensors and long term acclimation mediated by gradual changes in hypoxia-sensitive gene and protein expression. Both types of responses are graded in that their intensity increases in proportion to the severity of hypoxia until the limit of the response is reached. Integration of these responses is collectively responsible for overall acclimatization to hypoxia, and therefore, altitude. The acclimatization response enhances the ability of tissues to take up $O_2$ more efficiently for use by cellular processes, particularly respiration.

The body normally obtains $O_2$ from the air by taking it up from the lungs, where it passes by diffusion from the air spaces (alveoli) into capillaries and binds chemically to hemoglobin. Because the alveolar capillary membranes are not perfect gas exchangers, there is a slightly lower $PO_2$ in blood than in alveoli. The $O_2$ is then transported by the action of the circulation to the tissues, where it diffuses into the cells and is reduced to water by the mitochondria in the process of internal respiration. Thus, the mitochondria represent a sink for the disappearance of $O_2$, and the rate of $O_2$ consumption is determined by the needs of tissues for energy. The process by which molecular $O_2$ moves down its concentration gradient into the cell is known as the oxygen cascade (Fig. 15.2).

The body's rate of aerobic metabolism or $O_2$ consumption is determined by the needs of the tissues and met by the cardiac output and $O_2$ extraction according to the principle of conservation of matter. This relationship is described by the Fick equation:

$$VO_2 = C.O. \times (CaO_2 - CvO_2)$$

where $VO_2$ is the oxygen consumption of the body, C.O. is cardiac output, and $CaO_2$ and $CvO_2$ are arterial and venous $O_2$ contents, respectively. The oxygen delivery to tissues is given by the product of $C.O. \times CaO_2$. The difference between $CaO_2$ and $CvO_2$ is the amount of $O_2$ extracted, which, when divided by the amount originally present ($CaO_2$), provides the $O_2$ extraction ratio. The Fick equation shows that only two mechanisms are available to support an increase in metabolic demand—an increase in cardiac output and an increase in $O_2$ extraction. At rest only

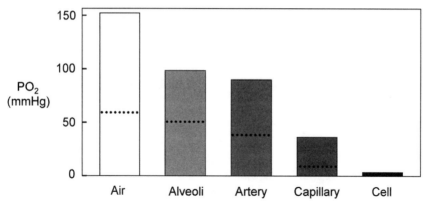

**Figure 15.2.** The oxygen cascade. The concentration of $O_2$ between air and the tissues progressively decreases, as shown by the fall in $PO_2$ from lungs (alveoli) to cells. In cells the mitochondria act as a sink for $O_2$ by converting it to water. Movement of $O_2$ from lungs to tissues, or $O_2$ transport, occurs by a series of physical processes: diffusion from the lungs to blood, chemical combination with hemoglobin, bulk (convective) delivery to tissue by the circulation, chemical release from hemoglobin, and diffusion into the tissues. Collectively, these processes are called a cascade. The dotted lines show the lower limit of the $PO_2$ needed in each compartment to supply enough $O_2$ for all cells to function properly.

about 25% of the $O_2$ available in the circulation is extracted, but when metabolic demand is increased, such as during heavy exercise, the heart and skeletal muscles can consume more than 75% of the $O_2$ available. Because cardiac output can increase fivefold and the $O_2$ extraction ratio threefold, a well-conditioned individual can increase $VO_2$ by fifteenfold (15 mets). Some highly trained elite athletes can increase cardiac output by more than seven-fold and achieve exercise values of 20 mets!

The maximum aerobic power ($VO_2$ max) of the human body is therefore set by two limits: maximum $O_2$ delivery and maximum $O_2$ extraction. Normally, the $O_2$ delivery limit is set primarily by cardiac output because arterial content ($CaO_2$) is constant during exercise. The limit of $O_2$ extraction is determined primarily from the number of working mitochondria. These two limits are closely matched in humans as well as throughout the animal kingdom; there does not seem to be an excess capacity in either part of the system (see Weibel, 1986). Therefore, the only way to increase the $VO_2$ max is to increase muscle mass, which increases the volume of mitochondria available to consume oxygen. Well-trained young athletes reach $VO_2$ max at 50 milliliters per kilogram per minute and elite athletes at 70 to 90 milliliters per minute per kilogram of body weight, depending on the type of exercise.

A relative lack of $O_2$ in the alveolar air at high altitude (hypoxia) always causes a greater lack of $O_2$ in the blood (hypoxemia) because $O_2$ always moves down its concentration gradient. Hypoxemia is compensated for by cardiovascular responses that maintain $O_2$ delivery by increasing blood flow and redistributing it to organs with the greatest $O_2$ needs. These responses are similar to those of exercise, although they are regulated by different mechanisms. It is important to note that hypoxemia decreases $CaO_2$, and therefore $VO_2$ max is very sensitive to high altitude. As hypoxemia progresses, such as during ascent, it produces cellular hypoxia, which leads to physiological adaptation to altitude.

The earliest and most pronounced adaptive effect of hypoxemia is that of increasing pulmonary ventilation. This hyperventilation raises the $PO_2$ in the lungs by washing metabolic $CO_2$ out of the alveoli more rapidly than is usual. This is a direct consequence of Dalton's law of partial pressures. Hyperventilation can decrease arterial and hence alveolar $PCO_2$ from a normal value of approximately 40 millimeters Hg at sea level to values as low as 7 to 10 millimeters Hg at the summit of Mt. Everest. Thus, hyperventilation allows $PO_2$ in the alveoli to fall more slowly as altitude increases. At 8000 meters, the predicted $PO_2$ in the lung stabilizes at about 35 millimeters Hg due to hyperventilation.

Almost everyone who ascends to high altitude for more than a few hours is helped by acclimatization. As mentioned, high-altitude acclimatization is manifest first by hyperventilation, which is caused by the effect of hypoxemia on a circulatory oxygen sensor called the carotid body, a small bundle of nerve cells located near the arch of the aorta that have chemoreceptor activity. The activity of carotid body cells is stimulated by hypoxemia, and impulses are transmitted to the respiratory centers in the brain to increase ventilation. This effect doubles the ventilation (and roughly halves the $PCO_2$) in someone acclimatized to an altitude of 15,000 feet (4,572 meters).

The extent of hyperventilation and thus the fall in $PCO_2$ at altitude is attributable to the degree of hypoxia and to individual factors that may have a genetic basis. People who increase ventilation more than average tolerate high-altitude exposure better than do those with a poor ventilation response. Thus, a strong respiratory response to hypoxia is a protective factor during ascent to altitude. The simplest way to think about this response is by the reciprocal relationship between $PCO_2$ and $PO_2$ in the lungs. The greater the decrease is in $PCO_2$, the higher the lung's $PO_2$ and the less severe the hypoxemia.

Elimination of extra $CO_2$ by ventilation also decreases hydrogen ion concentration (pH) in the blood, which produces alkalinity (respiratory alkalosis). Alkalosis increases the affinity of the hemoglobin molecule for $O_2$, thereby making it easier for red blood cells to obtain $O_2$ from lung capillaries. In normal blood hemoglobin is half-saturated with $O_2$ when $PO_2$ is approximately 27 millimeters Hg. The presence of alkalosis shifts the position of the hemoglobin saturation curve to the left and allows hemoglobin to be half-saturated at a lower $PO_2$. This effect makes it easier for hemoglobin to pick up $O_2$ in the lungs but harder to release it to the tissues. If the alkalosis persists it is compensated for in part by an increase in the concentration of a molecule called 2,3 diphosphoglycerate (2,3 DPG) in the red blood cells that restores the ability of hemoglobin to release $O_2$ more easily. These principles are illustrated in Figure 15.3.

Shortly after arriving at altitude the kidneys begin to compensate for the effects of hyperventilation by eliminating excess base from the blood in the form of bicarbonate ($HCO_3^-$) anion. This renal compensation shifts the $O_2$ dissociation curve back toward its normal position. However, at extreme altitudes blood alkalosis predominates, and the $O_2$ dissociation curve remains shifted to the left. In addition, the kidneys excrete salt and water, which decreases the plasma volume and helps increase hemoglobin concentration during the first two or three days of exposure to altitude.

An increase in hemoglobin concentration and number of circulating red blood cells (erythrocytosis) is a major feature of altitude acclimatization. Erythrocytosis develops gradually and requires many days for completion. The combined effect of erythrocytosis and the decrease in plasma volume at altitude is termed polycythemia. Polycythemia increases the capacity of blood to carry $O_2$ by as much as one-third depending upon the altitude and the extent of the individual's acclimatization response. Hemoglobin values, normally about 15 grams per deciliter, approach 20 grams per deciliter in people who have acclimatized fully to altitude.

Another important aspect of acclimatization is an increase in blood flow to the brain and heart to maintain their $O_2$ supply. In other tissues increases have been measured in number of capillaries per volume of tissue (capillary volume density) later in acclimatization. In skeletal muscle, however, the capillary effect appears to be due in large part to a decrease in muscle fiber size (atrophy) rather than to an actual increase in number of capillaries. This response differs from endurance training

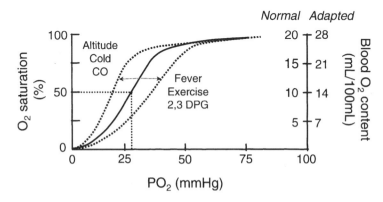

**Figure 15.3.** Oxygen dissociation curves for human hemoglobin in sea level– and altitude-adapted individuals. The blood oxygen content at any $PO_2$ is higher in adapted individuals due to an increase in blood hemoglobin concentration and the increase in the oxygen affinity of hemoglobin (curve is shifted to the left at altitude).

of the muscle, whereby both the capillary and mitochondrial volume density increase by responses regulated by cell growth factors.

There is evidence that chronic hypoxia increases the concentrations of some metabolic enzymes that produce energy adenosine triphosphate (ATP) in the cells. Enzymes that produce energy from glucose and in some cases even enzymes inside the mitochondria that are involved in the citric acid cycle and respiration, which depend on $O_2$ for their activities, have been found to increase with prolonged hypoxia. There are conflicting data, however, about the size and volume density of mitochondria after altitude acclimatization, and changes have been reported in both directions.

One reason for this discrepancy is a decrease in fluid in the interstitial and intracellular compartments. Loss of body fluids results from voluntary decreases in salt and water intake and increased renal salt and water excretion at altitude. These renal effects are mediated partly by endocrine mechanisms that decrease plasma volume and partly by mechanisms not entirely understood that involve reflex stimulation of the renal nerve by carotid body activity.

In acclimatization to moderate altitudes, weight loss stabilizes, and with time some actual increases may occur in the muscle cells' metabolic machinery. This makes sense in terms of the efficiency of mitochondrial $O_2$ extraction, but demonstrating this after acclimatization has been difficult experimentally. However, at extreme altitude, where deterioration and weight loss are problems, this kind of cellular adaptation is unlikely to be important.

Adequate acclimatization is absolutely essential to life at high altitude as well as to the success of climbing, skiing, and other physical pursuits. Without acclimatization the effects of hypoxia limit both physical and mental capacity in proportion to the severity of the stress. For example, people exposed acutely to the altitude of Mt. Everest without acclimatization remain conscious for only a minute or two, yet acclimatized mountaineers can follow their routes and gradually climb to the summit.

How long does complete acclimatization take and what limits the ability of acclimatized individuals to ascend to even greater altitude? The first part of the question is difficult to answer because altitude acclimatization is a series of complex responses that develop at different rates depending on the particular response, ascent rate, and altitude. The main responses of the cardiopulmonary system, including hyperventilation, commence almost immediately after arriving at altitude and develop fully in days to weeks. Other responses, including the increase in hemoglobin concentration, develop more gradually because they require actual changes in gene expression. Hemoglobin reaches its maximum concentration after several weeks at altitude, but this response is not simple, either, because hypoxia first decreases plasma volume, then increases red blood cell mass under the control of the renal hormone erythropoietin (EPO).

EPO is synthesized in the kidney and released in increased amounts into the circulation during anemia and hypoxia. It stimulates production of red blood cells by the bone marrow. The EPO response to hypoxia is controlled with a cellular pathway mediated by a protein complex known as HIF-1 (hypoxia-inducible factor-1). HIF-1 is present normally in many cells, but it is inactive because one of its proteins, the α subunit, is rapidly degraded when $O_2$ is present (Figure 15.4). However, in the presence of hypoxia the α subunit is stabilized, and HIF-1 complex is formed and transported (translocation) to the nucleus of the cell, where it binds to a specific region of the gene that encodes for EPO. This region of DNA, called the promoter, regulates the transcription of the EPO gene. Thus, HIF-1 is a transcription factor involved in the response to hypoxia.

It is notable that HIF-1 also regulates the expression of other genes that respond to hypoxia, such as vascular endothelial growth factor (VEGF), which stimulates the growth of new blood vessels (angiogenesis). Other transcription factors are also known to respond to hypoxia, and others have surely yet to be discovered. The time for complete molecular acclimatization on ascent to the response's limit (somewhere >8000 meters) is not known precisely but appears to be between six and ten weeks.

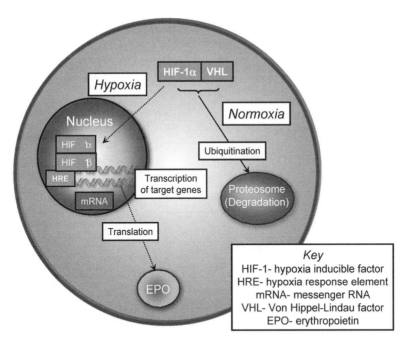

**Figure 15.4.** Production of erythropoietin (EPO) hormone in kidney cells is induced by hypoxia-inducible factor-1 (HIF-1). HIF-1 requires association (dimerization) of two peptide subunits, α and β, followed by nuclear translocation and DNA binding to activate transcription of the EPO gene. Normally, dimerization is prevented by the presence of Von Hippel-Lindau (VHL) factor. Hypoxia inhibits an enzyme (proline hydroxylase) needed for VHL to bind HIF-1α, thereby allowing it to associate with HIF-1β. EPO is synthesized and stimulates the production of erythrocytes by bone marrow.

Once a climber or visitor to altitude has become acclimatized, his or her ability to exercise at altitude is enhanced. This effect has led to efforts to improve athletic performance by training at altitude and by the use of EPO as a performance-enhancing hormone. The altitude strategy so far has little to recommend it because it is technically difficult and the chance of developing mountain sickness is significant. The doping strategy, based on injection of EPO to boost red blood cell mass, improves performance slightly for endurance athletes, but it is dangerous because it does not increase plasma volume and thus increases the viscosity of the blood. This increases the risk of clot formation. These problems have resulted in an international ban on the use of EPO by competitive athletes.

Complete acclimatization to altitude requires weeks, but some responses to hypoxia slowly dissipate, or "roll-off," with the time spent at constant altitude. For example, the respiratory response to hypoxia becomes less pronounced in almost everyone as time passes at altitude. The effects of acclimatization also appear to be fully reversed within a few weeks after descent.

It has been estimated that 150 million people live permanently at high altitude (above 2500 meters, or 8200 feet). Most of these permanent high-altitude residents, or highlanders, live in the Andes between 9000 and 15,000 feet, where the climate, depending on latitude, remains conducive to agriculture. However, a number of mines support small groups of permanent residents at altitudes as high as 17,500 feet. Some of these individuals ascend to 19,000 feet every day to work in a mine and return to the lower altitude at night.

Many highlanders appear to have acquired permanent adaptations to hypoxia. They respond differently to it than do people born at or near sea level (lowlanders). Permanent residents of high altitude show features of adaptation similar to the process of acclimatization in lowlanders, including increased ventilation and polycythemia. High-altitude natives have greater capacities for $O_2$ diffusion across the lung's capillaries both at rest and during exercise. In some studies increases in the density of the capillaries in the tissues also have been found compared to lowlanders.

The limit for permanent human adaptation to hypoxia appears to have been reached by Andean miners, who generally refuse to reside higher than 17,500 feet. These workers also appear to be self-selected for tolerance to hypoxia, because many others are unable to acclimatize to permanent life at such high altitudes. The miners are supplied with food from below, and the women descend to approximately 12,000 feet to give birth to children.

The issue of human infertility at high altitude has been an interesting and controversial topic for at least 500 years. The logical basis for an altitude limit for bearing children is lethal fetal hypoxia, but little rigorous scientific data have been collected on the problem. According to L. G. C. E. Pugh, writing in 1965, when Spanish conquistadores founded the imperial city of Potosí at 13,000 feet (4000 meters) in the sixteenth century in what is current day Bolivia, Spanish women had to descend to lower altitudes to bear children, while women of the indigenous population had no difficulty. The first child of Spanish parents to survive infancy was not born for more than fifty years and was known as the miracle of St. Nicholas Tolentino (Pugh, 1965).

After many generations, Spanish women apparently were able to bear children at Potosí, but eventually the capital was moved to Lima partly because of the low fertility of livestock. If the story is true, recovery of fertility in female descendents of Spanish women in Potosí remains unexplained. L. G. Pugh speculated, perhaps because the time seems too short for natural selection to have occurred, this was due to intermarriage. The lack of Spanish fecundity implies that fertility was restored by indigenous women unless it happened because of undiscovered factors in the gene pool of native males.

## High-Altitude Illnesses

Medical problems at high altitude generally occur in individuals who ascend too quickly to acclimatize adequately. The terms *altitude illness* and *mountain sickness* describe the continuum of medical syndromes of prolonged exposure to hypoxia. In most cases the effects of acclimatization cause these symptoms to subside within a few days. It is convenient to classify altitude illness by the original scheme proposed by Thomas Ravenhill in 1913. A British Royal Army physician, he described the basic types of altitude illness in his classic paper "Some Experience of Mountain Sickness in the Andes." These are acute mountain sickness (AMS), high-altitude pulmonary edema (HAPE), and high-altitude cerebral edema (HACE).

AMS occurs after rapid ascent to 2400 meters (about 8000 feet) or above. It develops gradually within six to twelve hours after arrival at altitude and usually subsides within two or three days if no further ascent is undertaken. AMS is invariably accompanied by symptoms of headache, and at least one of the following: dizziness fatigue, sleeplessness, nausea, vomiting, and loss of appetite. Sleep disturbances, which are common at altitude, are more severe in AMS.

HAPE is a serious and potentially fatal form of altitude illness described by Ravenhill as "puna of the cardiac type." It is often preceded by symptoms of AMS and usually occurs in the first two or three days after ascent to altitudes above 2500 meters. It is seen in less than 2% of people who follow recommended rates of ascent. These patients develop chest congestion and shortness of breath (dyspnea) with exercise that progresses to dyspnea at rest, particularly at night. Dry cough, also common at altitude, is more pronounced in HAPE and may progress to cough with pink, frothy sputum, indicating the onset of pulmonary edema. Pulmonary edema begins as a patchy leakage of fluid from the capillaries into the airspaces of the lungs, but this may progress rapidly and become diffuse, thereby causing worsening of hypoxemia. Edema of the face, hands, and feet may also be present.

HAPE is more common in men than in women, and it is frequently associated with exercise. Good fitness does not protect against HAPE, but poor fitness is a risk factor. Sensitive individuals have been described who are prone to repeated bouts of HAPE. Recurrent episodes generally show unique distributions of edema suggesting that underlying structural abnormalities of the lungs are not involved in its pathogenesis. Obese individuals and those with pre-existing lung diseases are at increased risk for HAPE.

HACE, or "puna of a nervous type," is a rare and potentially fatal altitude illness. HACE is preceded by symptoms of AMS, and it is most common at altitudes above 3500 meters. The symptoms of HACE are variable, but most patients show clumsiness and an unsteady gait (ataxia). They may have personality changes, confusion, somnolence, seizures, and hallucinations. If HACE is suspected, the patient should be taken immediately to a lower altitude. Otherwise coma and death may soon follow.

The physiological responses to altitude become more pronounced above 3500 meters as a result of the shape of the $O_2$ dissociation curve of hemoglobin. This curve is sigmoid and slopes steeply downward at an arterial $PO_2$ of 60 millimeters Hg; thus, the amount of $O_2$ in the blood falls in direct proportion to the fall in $PO_2$. For healthy people this occurs at altitudes above 10,000 feet. As arterial $PO_2$ declines to less than 55 millimeters Hg, it produces two important but opposing responses, an increase in pulmonary artery pressure caused by hypoxic constriction of lung blood vessels and an increase in cerebral blood flow (CBF) caused by dilation of brain blood vessels.

Hypoxic pulmonary vasoconstriction (HPV) maintains matching of local ventilation with local blood flow in the lungs. However, when hypoxia involves the entirety of both lungs, right-side cardiovascular pressures may become so high as to stress the lung's small blood vessels and capillaries and allow them to leak. It is also possible that hypoxia directly causes the capillary lining (endothelium) of the lung to become more permeable to protein and fluid, a condition called noncardiogenic pulmonary edema. Either or both of these effects could explain the protein-rich edema fluid found in the lungs of patients with HAPE. Migration of inflammatory cells to the lung also has been postulated to play a role in HAPE, but evidence for this remains limited.

In contrast to HAPE, HACE is related to vasodilation of blood vessels in the brain, which increases cerebral blood flow, and opposes the effects of hyperventilation. There are also inhibitory effects of low $PO_2$ on biochemical processes required to maintain the integrity of the blood–brain barrier (BBB). The BBB normally prevents free passage of solutes from the circulation into cerebral tissue. Disruption of this barrier by high hydrostatic or osmotic pressures, or by biochemical events promotes swelling (edema) of brain tissue. This edema, if progressive, may raise the intracranial pressure because the skull is a rigid container. High intracranial pressure may compromise blood flow to the brain because it decreases cerebral perfusion pressure, or the difference between average arterial pressure and intracranial pressure. In patients who die from HACE, however, the pattern of brain edema tends to be heterogeneous, with microscopic foci of capillary injury and red blood cell and platelet aggregates present. These findings suggest a complex basis for the etiology of HACE.

The most rational approach to altitude illness is to avoid it by gradual ascent and to stop ascent or descend if symptoms appear. Ideally, the rate of ascent should be approximately 1000 feet per day between 8000 and 10,000 feet and 500 feet per day for altitudes above 10,000 feet. Other ascent strategies also have been recommended based on spending each night at the lowest altitude possible.

If slow ascent is not practical, the incidence of AMS can be reduced by prophylactic treatment with the mild diuretic drug acetazolamide. Acetazolamide inhibits an enzyme, carbonic anhydrase, that reversibly converts $CO_2$ to bicarbonate. Acetazolamide increases bicarbonate excretion by the kidneys and lessens respiratory alkalosis. It is administered at a dose of 250 milligrams twice daily or 500 milligrams of the slow release preparation once daily beginning the day before ascent to 2500 meters or above. It should be continued until the day after maximum ascent or until acclimatization is complete. Acetazolamide has unpleasant side effects, such as interfering with the taste of carbonated beverages and may cause reactions in individuals allergic to sulfa drugs. The use of acetazolamide minimizes dehydration and potassium depletion. It is not the diuretic effect that is important, because liberal water intake facilitates bicarbonate excretion and prevents dehydration.

The only definitive treatment for altitude illness is descent to lower altitude. The basic principle when symptoms appear is to stop climbing and rest until acclimatization occurs. Ascent to higher sleeping altitude is contraindicated in the presence of symptoms of AMS. If the symptoms worsen over twenty-four hours or do not improve after seventy-two hours, descent is necessary. Immediate descent is required if signs or symptoms of HACE or HAPE develop. If there is doubt about symptom progression, the patient should be taken to an altitude below that where symptoms first appeared. Supplemental $O_2$ may be lifesaving but may not be available in remote locations.

A chronic form of mountain sickness occurs in people who have lived at very high altitudes for prolonged periods of time. Such highlanders have a lower respiratory response to hypoxia and lower ventilation at altitude than do residents at sea level (Monge, 1943). Occasionally, these individuals develop chronic mountain sickness (known as Monge's disease) characterized by an exaggerated increase in the pressure of the pulmonary circulation that leads to failure of the right side of the heart (cor pulmonale). Other physiological responses of chronic mountain sickness include pronounced polycythemia, in which the hemoglobin concentration may increase to as much as 28 grams per deciliter. Ventilation decreases because of loss of sensitivity of the respiratory center to hypoxia, which leads to elevated $PCO_2$, and worsens hypoxemia. Most patients with chronic mountain sickness have shortness of breath, cough, rapid heartbeat, and headache. The skin, particularly of fingers and toes, is often cyanotic (blue) in color. Neurological changes such as weakness, lethargy, and stupor may also occur. The only effective therapy is to alleviate hypoxemia by moving the affected individual to lower altitude. Chronic mountain sickness is not limited to humans; a similar syndrome, Brisket disease, has been described in cattle that live at high altitude.

## The Zone of Death

What causes death at high altitude? This question, hotly debated among physiologists for decades, has long been settled. Hypoxia, the major cause of altitude-related illness, is also the factor that limits survival. For many years mountaineers have recognized the deterioration in general physical condition, including weight loss, that develops during sustained exposure to extremely high altitude. This condition appears to differ from any of the well-defined high-altitude diseases. High-altitude deterioration comes about soon after one enters the "zone of death" above 8000 meters, and seasoned climbers make every effort to limit the time they spend above this altitude. Unfortunately, the higher one climbs, the smaller is the ascent that can be made each day.

At 8000 meters, the $PO_2$ in the atmosphere is approximately 56 millimeters Hg, which, in a tube of blood, would allow 88% of normal human hemoglobin to be saturated with oxygen. This $PO_2$ would not even require much acclimatization, but, as mentioned earlier, $PO_2$ in the air spaces, or alveoli, of the lungs is considerably lower than that in inspired air because $O_2$ is diluted by $CO_2$ and water vapor in accordance with Dalton's law. To compute the amount of $O_2$ that reaches the alveoli, the partial pressure of water vapor at body temperature (47 millimeters Hg) and metabolic $CO_2$ must be subtracted from the composition of the inspired air that enters the lungs. This is done using the alveolar air equation:

$$PAO_2 = FIO_2 \, (P_b - PH_2O) - \frac{P_aCO_2}{RQ}$$

where $PAO_2$ is the alveolar $PO_2$, $FIO_2$ is the fractional expression of inspired $O_2$ (0.21), $P_b$ is barometric pressure, $PH_2O$ is water vapor pressure (47 millimeters Hg), and $PACO_2/RQ$ is the alveolar $PCO_2$ divided by the respiratory quotient.

At sea level inspired air has a barometric pressure of 760 millimeters Hg, which, after subtracting 47, leaves 713 millimeters Hg. The $PO_2$ of air as it passes into the trachea is 713 x 0.21, or 150 millimeters Hg. The partial pressure of $CO_2$ entering the alveoli from the blood must then be subtracted from the total. The amount of $CO_2$ that normally reaches the alveoli has a partial pressure of 40 millimeters Hg, which, at an RQ of 0.8, dilutes the $PO_2$ in the alveoli by about a third. Thus, the $PO_2$ in the alveoli of healthy young adults is normally 100 millimeters Hg. The $PO_2$ in the arterial blood is only about 5 to 10 millimeters Hg less than the alveolar value because a normal lung is a highly efficient gas exchanger.

During an ascent to 8000 meters, if the $PCO_2$ in the alveoli remained at 40 millimeters Hg, the alveolar $PO_2$ would be only 6 millimeters Hg, and the arterial value would be nearly zero! This would happen at 8000 meters because inspired air has a $PO_2$ of (267 − 47 millimeters Hg) x 0.21, or 46 millimeters Hg. Fortunately, hyperventilation prevents this, as will be illustrated below, by lowering the $PCO_2$ to vigorously defend arterial $PO_2$. In general, this mechanism prevents arterial $PO_2$ from falling below about 30 millimeters Hg.

The most intriguing data that indicates that the physiological limit of human tolerance to altitude lies just above 8000 meters comes from the death rates of climbers on Earth's fourteen mountain peaks at or above this altitude (Table 15.1). Between 1950 and 2000 more than 600 climbers on these fourteen peaks were killed. One way the problem has been evaluated is by examining the number of deaths of climbers that occur while descending after having reached an 8000-meter summit. These climbers, having been extended to the maximum, are often exhausted, which makes them more vulnerable to falls and other accidents. On average, the probability of dying during descent from above 8000 meters is about 1

**Table 15.1.** Mountain Peaks over 8000 Meters

| PEAK | ALTITUDE (m) | AMBIENT $PO_2$ (mmHg)* | LOCATION |
| --- | --- | --- | --- |
| Everest | 8850 | 49.3 | Nepal/ Tibet |
| K2 | 8611 | 51.0 | Pakistan |
| Kangchenjunga | 8586 | 51.4 | Nepal/India |
| Lhotse | 8516 | 51.8 | Nepal/Tibet |
| Makalu | 8463 | 52.3 | Nepal/Tibet |
| Cho Oyu | 8201 | 54.3 | Nepal/Tibet |
| Dhaulagiri | 8167 | 54.5 | Nepal |
| Manaslu | 8163 | 54.5 | Nepal |
| Nanga Parbat | 8126 | 55.0 | Pakistan |
| Annapurna I | 8091 | 55.2 | Nepal |
| Gasherbrum I | 8068 | 55.6 | Pakistan/China |
| Broad Peak | 8047 | 55.8 | Pakistan/China |
| Gasherbrum II | 8035 | 55.9 | Pakistan/China |
| Shishapangma | 8027 | 56.0 | Tibet |

*Ambient $PO_2$ is calculated from the standard atmosphere and is only approximate because of effects of latitude (see West, 1999).

in 30, but risk of death also correlates with technical difficulty and the altitude of the peak. For instance, the death rate on descent from Everest is 1 in 12, while on the world's second-highest peak, K2, it is 1 in 6. K2, known indigenously as Chomori and located at latitude 35°N in the Karakoram Range of northeast Pakistan, is nearly 8° north of Everest and more challenging technically (Figure 15.5). On peaks of 8000 to 8200 meters (except Annapurna I, which is also technically difficult) the death rate is 1 in 100 to 1 in 200.

Can a survival effect be discerned that is due to hypoxia rather than technical factors? Because the 8000-meter plus peaks differ in altitude by only a little more than 800 meters, the difference in inspired $PO_2$ is just 7 millimeters Hg. It has been noted that the death rate on descent from K2 is considerably lower for climbers who breathe extra $O_2$ throughout the climb, which is physiologically equivalent to reducing the altitude (Huey and Eguskitra, 2000). This observation does not exclude technical contributions, but technical rather than physiological explanations seem less likely because only climbers who had already attained the summit were included in the analysis. There may also be other benefits of oxygen, such

**Figure 15.5.** K2 (Chomori) is the world's second-highest peak and is also one of the most technically challenging and deadliest climbs for mountaineers. The death rate of climbers descending from K2's summit is five times the average for peaks higher than 8000 meters and more than twice that of Mt. Everest. (Copyright Chris Warner, Earth Treks Climbing Center, reproduced with permission.)

as protecting the body from the cold by allowing the climber to shiver more vigorously at high altitude.

## Limits of Human Ascent to High Altitude

How high can acclimatized individuals ascend, and what determines the limit? The summit of Mt. Everest, at 8850 meters, seem close to this limit because although ascent to the summit with oxygen occurred in 1953, it was not climbed without oxygen for another twenty-five years. Three years later, during the 1981 American Medical Research Expedition to Everest (AMREE) under the direction of Dr. John West, actual measurements of the respiratory gases of climbers were made (see West, 1998). At a barometric pressure of 253 millimeters Hg and an ambient $PO_2$ of 53 millimeters Hg, the alveolar $PO_2$ was approximately 37 millimeters Hg (normal at sea level is 90 to 100 millimeters Hg) and the alveolar $PCO_2$ was 7.5 millimeters Hg (40 millimeters Hg is normal).

The surprise in this information, assuming the measurements were correct, is that the $PCO_2$ in these well-acclimatized climbers was unexpectedly low. This enabled the $PO_2$ to be slightly higher than that found on previous expeditions. In earlier studies alveolar $PO_2$ values of 33 to 40 millimeters Hg, and $PCO_2$ values of 14 to 21 millimeters Hg had been measured in climbers at 288 to 347 millimeters Hg (well below the summit of Everest). Thus, it appears that climbers at extreme altitude maintain their arterial $PO_2$ above 30 millimeters Hg by increasing the extent of hyperventilation. The benefit to climbers is that it provides more $O_2$ for physical work, although the cost is to increase the work of breathing, which requires greater consumption of $O_2$ by the respiratory muscles. In any event, this trade-off effect indicates that the body is attempting to stave off an approaching physiological limit presumably related to the minimum $O_2$ consumption necessary to continue to climb.

Using numerical methods to compute the maximum $O_2$ consumption of climbers on the AMREE, West estimated it to be just over 1 liter per minute, which amounts to less than one-fourth of the sea level value (see Fig. 15.6). Under these conditions a climber would be able to walk on level ground at a deliberate pace. If the ascent continued, the climber would probably no longer be able walk when the maximum $O_2$ consumption fell to 10% of its sea level value because of inability to breathe and walk simultaneously. Extrapolation from the curve in Figure 15.6 to the 10% value indicates that inspired $PO_2$ would be approximately 40 millimeters Hg and barometric pressure about 240 millimeters Hg when the limit is reached. At the latitude of Mt. Everest, these conditions would occur at an altitude of approximately 9250 meters (30,078 feet).

Where does this limit place humans with respect to hypoxia-tolerant species in the rest of the animal kingdom? Compared to other mammals, this is difficult to

180    THE BIOLOGY OF HUMAN SURVIVAL

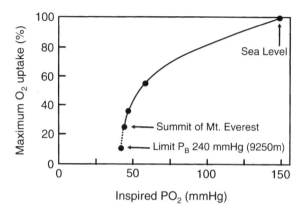

American Medical Research Expedition to Everest (AMREE 1981)

**Figure 15.6.** Maximum oxygen consumption of the body as a function of inspired $PO_2$ on ascent to high altitude. Oxygen consumption (Y-axis) is expressed as a percentage of the sea level value as altitude increases ($PO_2$ on X-axis). The calculated limit is reached at 9250 meters. (Data adapted from West, John B. 1998. *High Life: A History of High Altitude Physiology and Medicine.* New York: Oxford University Press, Chapter 11.)

say because very few mammals are exposed naturally to extreme hypoxia for long periods of time. Mammals rarely reside permanently at altitudes above 5000 meters. Among other vertebrates, some turtles can tolerate a complete lack of $O_2$ for many hours, but these animals are cold-blooded, at rest, and their brain $O_2$ requirements are shut down in anoxia. Thus, the true high-altitude champions are clearly the birds. For instance, the bar-headed goose migrates annually over the Himalayas at altitudes between 10,000 and 11,000 meters (33,333 to 36,667 feet) from India to its breeding grounds in Central Asia. In other words, it can fly comfortably at least a mile above the human altitude limit. At 11,000 meters the $PO_2$ of the air that enters the lungs of a bar-headed goose in flight is less than 35 millimeters Hg. When it comes to exercise at high altitude, birds have several important advantages over humans. The avian lung is designed with a more efficient countercurrent gas exchange mechanism to extract $O_2$, avian hemoglobin has a higher $O_2$ affinity than that of acclimatized humans, and the bird is insensitive to decreases in blood $CO_2$ content; it can hyperventilate without restricting cerebral blood flow to the point at which it passes out.

# 16

# Into the Wild Blue Yonder

The Earth's gaseous atmosphere is an essential part of the biosphere of the planet. It provides the $O_2$ required by most forms of life, a stable thermal environment, and protection from the harmful effects of cosmic radiation. Only the very closest atmospheric region to the surface of the Earth, which extends from sea level to an altitude of about 16 kilometers (10 miles), can support aerobic life. This is a mere 3% of the complete atmosphere of the planet. The thinness of the air above this life-sustaining envelope, however, has not prevented humans from testing their limits against it.

## The International Standard Atmosphere

By convention the atmosphere is divided into a series of layers, or concentric shells, characterized primarily by their thermal properties. This structural arrangement, known as the International Standard Atmosphere, divides the atmosphere into five regions: the troposphere, the stratosphere, the mesosphere, the ionosphere (thermosphere), and the exosphere (Fig. 16.1). These regions are separated by theoretical boundaries known as pauses.

The atmosphere near the surface, the troposphere, which extends from sea level to an altitude of 16 kilometers, can support life. The air in the troposphere be-

182    THE BIOLOGY OF HUMAN SURVIVAL

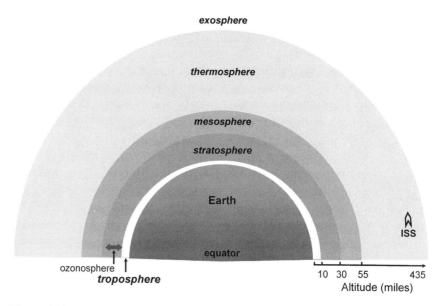

**Figure 16.1.** Composition of the International Standard Atmosphere. Animal life is possible only in the troposphere, within 14 kilometers (9 miles) of the surface.

comes progressively colder as the altitude increases. The rate of cooling of the atmosphere, called the temperature lapse rate, averages 2°C per 300 meters (1000 feet) of altitude. The air temperature at the boundary between the troposphere and the stratosphere (the tropopause) varies from −55°C to −80°C depending on latitude. Over the poles the tropopause is actually warmer than it is at the equator because the greater intensity of sunlight at the equator causes the air layer to rise from the surface more than twice as far as it does at the poles. In addition, the altitude of the tropopause averages 7800 meters (26,000 feet) at the poles and 16,800 meters (56,000 feet) at the equator.

The second layer of the atmosphere, the stratosphere, extends to 50 kilometers above the surface. The stratosphere is a cold, inhospitable place. The temperature of the stratosphere actually rises from about −55°C at the tropopause to 0°C at the stratopause, where stratosphere meets mesosphere. The increase in temperature is due to exothermic dissolution of ozone within the stratosphere some 20 kilometers to 40 kilometers above the surface. This region of the stratosphere, known as the ozonosphere, contains the critical layer of ozone that protects life from the damaging effects of solar ultraviolet radiation.

The mesosphere, or middle portion of the atmosphere, is located between 50 kilometers to 85 kilometers above sea level. The temperature of the mesosphere plummets to −100°C as it approaches the ionosphere (or thermosphere), which extends from 85 kilometers to 700 kilometers above the surface, where it merges gradually into the exosphere, or the boundary between the Earth's atmosphere and interplanetary space. The air in the thermosphere is so thin that it does not create appreciable friction, but layers of ionized particles reflect heat from the planet back toward the surface. The temperature of the thermosphere depends on solar activity and may exceed 1500°C during the daytime. This remarkably high atmospheric temperature, however, does not have the same meaning it would have near sea level because heat transfer in the widely dispersed molecules of the thermosphere occurs almost exclusively by radiant, not by conductive–convective, exchange.

In the vicinity of the equator 1000 to 25,000 kilometers above Earth, a dense band of ionized particles, the Van Allen belt, surrounds the planet. Beyond the Van Allen belt, interplanetary space is almost empty except for scattered hydrogen atoms and occasional solid objects ranging in size from dust particles to asteroids. Because of the molecular void, there is no atmospheric temperature, and heat transfer by convection does not exist. The temperature of interplanetary dust in Earth's neighborhood of the solar system is approximately –170°C owing to radiant energy flux from the Sun.

## Human Visitation to the Stratosphere

Human visitors have frequented the stratosphere since the early part of the twentieth century, but survival there requires more than supplemental oxygen. It is possible only in pressurized cabins for reasons that will become apparent below. This problem was recognized and solved by Auguste Piccard (1884–1962), a Swiss-born physicist who invented the pressure capsule. The scientific efforts of Piccard and his twin brother, Jean, are directly responsible for opening up the stratosphere to human travel.

In 1931 Auguste Piccard made the first flight into the Earth's stratosphere, rising in a sealed aluminum gondola suspended beneath a hydrogen-filled balloon to an altitude of 15,797 meters (51,793 feet, or 9.8 miles). The following year he ascended to 16,507 meters (54,120 feet, or 10.25 miles). The barometric pressure during Piccard's flights fell from 760 millimeters Hg (1000 millibars) to as low as 66 millimeters Hg (87 millibars) at an altitude 10 miles above sea level. Oxygen was supplied by evaporation of liquid oxygen, and $CO_2$ was removed from the cabin using sodalime. During these flights into the stratosphere, he recorded air temperatures ranging from –55°C to –60°C (–67°F to –76°F).

Writing in 1933 of his record-breaking ascent into the stratosphere, Piccard reported the first potential aviation disaster averted by the use of liquid oxygen. To pressurize the cabin during ascent, Piccard and his young companion, Paul Kipfer, had to close a one-inch instrument hole in the gondola. Before they could plug the hole, the balloon had reached 15,000 feet, with air still whistling merrily out. The altimeters indicated that the internal and external pressures were nearly identical. Pouring small amounts of liquid oxygen on the floor to evaporate, the men remained alert while they sealed the leak. After twenty-eight minutes the gondola was finally airtight but had already risen to an altitude of 9.65 miles!

Auguste Piccard's innovative pressurized gondola and his spectacular altitude adventures contributed immensely to the aeronautical knowledge needed to put humans safely into space. After World War II, however, Piccard turned his attention to undersea research with submersibles. In a bathyscaphe of his own design, he made several deep descents into the Mediterranean Sea with his son, Jacques. Like a submarine, the bathyscaphe has a hard pressure-resistant hull that protects its occupants from exposure to the pressure of the ocean depths. Piccard also designed the Italian-built submersible *Trieste* (Fig. 16.2), which was operated by the U.S. Navy and on January 23, 1960, descended to 35,802 feet (10,912 meters) in the Marianas Trench, the deepest part of the ocean floor, in the Challenger Deep near the Philippines. *Trieste*, carrying two men, Jacques Piccard and Donald Walsh, took nearly five hours to sink to the ocean floor.

The highest altitude for a balloon flight was recorded on May 4, 1961, by two U.S. Navy officers, Malcolm Ross and Victor Prather, when they reached 34,679

**Figure 16.2.** The *Trieste*. The bathyscaphe, operated by the U.S. Navy was designed by Auguste Piccard to resist the greatest depths of the Pacific Ocean. *Trieste* made a descent to the deepest point—35,802 ft (10,912 m) in the Marianas Trench on January 23, 1960 with two men in the gondola suspended beneath the large hull, which was filled with gasoline and lead shot to regulate buoyancy. (U.S. Naval Historical Society, photograph #96799)

meters (113,739 feet, or 21.5 miles) in the capsule of their hydrogen-filled polyethylene balloon. The atmospheric pressure at that altitude is approximately 0.25 millimeters Hg (3 millibars), roughly half the atmospheric pressure on the surface of Mars. Their balloon descended into the Gulf of Mexico, where the unfortunate Prather's pressure suit filled with water, and he drowned.

The need for supplemental oxygen during high-altitude balloon flight was made clear more than 125 years ago by the flights of balloonists who died on their adventures. The cause of death was hypoxia. Perhaps the most notorious incident was the ascent of the *Zenith* in France in 1875. The *Zenith* carried three men, the famous balloon pilot Theodore Sivel, an engineer, Joseph Corce-Spinelli, and a passenger, Gaston Tissandier. The men were aware of the dangers of hypoxia, having experienced it in Paul Bert's altitude chamber in Paris. On Bert's advice, the *Zenith* was equipped with bags of $O_2$–enriched air that the men could breathe from intermittently. During the flight, however, all three men became disoriented and lost consciousness as the balloon ascended to approximately 28,000 feet; only Tissandier survived the misadventure. The reason the men died was lack of acclimatization to altitude, for they ascended from near sea level to 28,000 feet in about two hours. Recall that 103 years later Messner and Habeler were able to climb Mt. Everest (29,028 feet) without extra $O_2$, but required several weeks.

Without supplemental $O_2$, the maximum altitude that can be tolerated before loss of consciousness even by humans acclimated to hypoxia is roughly 33,000 feet. The use of pure $O_2$ breathing gas can raise this limit to about 48,000 feet. At that altitude atmospheric pressure is approximately 94 millimeters Hg. Because the pressure of water vapor in the lungs is 47 millimeters Hg, the maximum partial pressure that can be achieved for the other gases in the lungs is also 47 millimeters Hg. Nevertheless, Dalton's law of partial pressures and the need to eliminate $CO_2$ produced by metabolism substantially reduces the $PO_2$ in the lungs. As noted in Chapter 15, the $PCO_2$ in normal lungs is 40 millimeters Hg, but extreme hyperventilation can decrease this value to less than 10 millimeters Hg. The maximum $PO_2$ that could be sustained in the lungs under these conditions is 47 – 10, or 37 millimeters Hg. Even with near-perfect exchange of $O_2$ between airspaces and capillaries in the lungs, the $PO_2$ in the arterial blood will approach 30 millimeters Hg, the approximate limit for maintaining function of the human brain after acclimatization to hypoxia.

## Depressurization Accidents

Today's commercial jet liners routinely crisscross the surface of the Earth by flying through the upper troposphere and lower stratosphere. Air travelers in supersonic transports fly as high as 60,000 feet, comfortably protected from the lethal effects of stratospheric hypoxia by pressurized cabins that maintain internal altitude equivalents of 8000 feet or lower. If an aircraft cabin depressurizes at such high altitudes, however, the consequences to the occupants can be rapidly lethal.

A notorious incident of this type occurred on October 25, 1999 when a Learjet crashed in South Dakota, killing the two crewmembers and four passengers, including professional golfer Payne Stewart. The aircraft had taken off from Orlando, Florida, at about 9:20 A.M. Eastern Daylight Time (EDT) headed for Dallas, Texas. North of Gainesville, Florida, air traffic control cleared the airplane to an altitude of 39,000 feet, when radio contact with the flight was lost. Shortly after noon Central Daylight Time (CDT) the jet crashed near Aberdeen, South Dakota. As it flew off course the Learjet was intercepted by U.S. Air Force and Air National Guard aircraft. The military pilots who flew beside the wayward jet reported they could not see into the cabin because its windshields were clouded by condensation. They noticed no external problems, and the aircraft was in normal flight. They then watched helplessly as the jet spiraled to the ground and crashed in an open field, killing all six occupants and destroying the airplane.

During the accident investigation, the National Transportation Safety Board (NTSB) determined that the flight crew had contacted the Jacksonville Air Traffic Control Center at 9:27 EDT and reported climbing through an altitude of 23,000 feet. The controller instructed the jet to climb and maintain an altitude of 39,000 feet, which was acknowledged by the crew. This was the last radio transmission received from the Learjet. At the time the first officer's speech was normal and appropriate, and later analysis of the recordings indicated she had not been using an oxygen mask microphone for her transmissions. The first sign of a problem was the crew's failure between 9:33 and 9:39 to respond to repeated inquiries as the airplane was climbing through about 36,000 feet, indicating that sometime during this six-minute period the crewmembers had been incapacitated. As the flight continued, the jet deviated from its course and failed to level off at its assigned altitude. Over the next four hours, there was no sign of crew activity. The airplane ascended to 48,900 feet and eventually ran out of fuel.

The cabin altitude audio warning sounded continuously during the final thirty minutes of flight (the only portion recorded), indicating that the airplane had lost cabin pressure. Condensation or frost on the windshield suggested loss of inflow air (bleed air) to the cabin because the bleed air supply to the cockpit provides a flow of warm air at the windshield and prevents condensation regardless of cabin temperature or pressure. A constant supply of bleed air to the windshield would have allowed it to remain relatively clear even after depressurization from a breach of the cabin, whereas depressurization by a loss of cabin air inflow would have generated frost. The jet therefore most likely had no air inflow into the cabin. There was also no other reason for the crew to have become incapacitated.

The NTSB performed computer simulations to try to understand the Learjet's cabin functions during ascent. The most insightful simulation assumed that cabin air was lost at altitudes of 30,000, 35,000 or 40,000 feet because of closure or failure of the cockpit air inflow valve. The simulation predicted cabin altitude would ascend to 25,000 feet in approximately 2.5 minutes and approach flight altitude within four to five minutes of valve failure. Meanwhile, in its investigation the FAA found that the emergency procedures for the Learjet Model 35 did not require the flight crew to don $O_2$ masks immediately upon activation of the cabin altitude audio warning. This oversight probably allowed the flight crew to troubleshoot the pressurization system before donning $O_2$ masks, which may have led to their incapacitation. Therefore, $PO_2$ in the cabin soon after depressurization was too low to support consciousness. Loss of consciousness in nonacclimatized individuals at an altitude of 25,000 feet or above would occur in two minutes or less (see again Fig. 13.2).

If the pilots had received $O_2$ from the airplane's emergency system, they likely would have responded appropriately to the depressurization by descending to a safe altitude. It is therefore almost certain that after depressurization the pilots did not receive supplemental $O_2$ in time or at an adequate concentration to avoid hypoxia. Examination of the wreckage of the airplane indicated that the pressure regulator valve on the oxygen bottle had been open on the flight, and the bottle was empty. The cockpit $O_2$ pressure gauge showed "witness marks" caused by the ground impact that indicated a lack of pressure in the $O_2$ bottle, but exactly when the bottle was exhausted during the four-hour flight is unknown. Evidence at the crash site suggested both crew masks were connected to the airplane's $O_2$ supply at the time of impact. In addition, both mask microphones were plugged into the crew microphone jacks. Therefore, $O_2$ should have been available to the pilots' masks if the oxygen bottle had been charged properly. In any event, the accident was caused by a classic double failure because it required both loss of cabin pressure and failure of the flight crew to receive adequate oxygen.

Several possibilities were evaluated for the crewmembers' failure to receive supplemental oxygen, including failure of the pilots to don masks rapidly enough after loss of cabin pressure, inadequate supply of oxygen, and improper filling of the supply bottle. That $O_2$ was insufficient to support the crew while they attempted to restore cabin pressure is a real possibility because even a full bottle on the aircraft would have supported six passengers for only about 10 minutes. If the bottle was not full or if it had been incorrectly charged with air, unconsciousness would have occurred sooner. The NTSB conclusion was that the cause of the accident, for undetermined reasons, was probably "incapacitation of the crewmembers as a result of their failure to receive supplemental oxygen following a loss of cabin pressurization."

The use of $O_2$ as a remedy for cabin depressurization, even if it is immediately available, does have limitations. For example, at 40,000 feet, where the barometric pressure is only 140 millimeters Hg, alveolar $PO_2$ can fall as low as 53 millimeters Hg even while breathing 100% $O_2$ (if the individual does not hyperventilate). If there is no cabin pressure above 40,000 feet, additional $O_2$ can be forced into the lungs using positive pressure breathing masks. The application of positive pressure to the lungs increases $PO_2$ in the alveoli by an amount proportional to the pressure because 100% $O_2$ is breathed; it also tends to stimulate ventilation and lower $PCO_2$, which improves oxygenation. A tight-fitting positive pressure mask can raise the high-altitude flight limit to roughly 52,000 feet. However, the application of large amounts of positive pressure to the chest has deleterious effects on cardiopulmonary function, and, in practical terms, the limit of a positive pressure mask is about 60 millimeters Hg.

Adequate cabin pressurization is still required to fly through most of the stratosphere even when pure $O_2$ and positive pressure are used to support respiration. This absolute requirement for cabin pressure speaks to one of the most curious and dramatic limitations to human survival, which is imposed by the physics of liquids and gases. Therefore, adaptation of the human body to it will remain forever impossible.

## The Armstrong Line

To help understand the nature of this pressure limit, it is useful to recall what everyone knows who has camped overnight in the high mountains and boiled water for coffee the next morning: it is much easier to boil water at high altitude. Water at high altitude boils at a lower temperature than does water at sea level because as atmospheric pressure falls the energy needed for water to evaporate also falls. At the summit of Pikes Peak (14,110 feet), the atmospheric pressure is 445 millimeters Hg, and water boils at about 88° C (190° F) instead of 100°C (212° F). At an altitude of 62,800 feet (19.14 kilometers, or 11.4 miles), the atmospheric pressure is 47 millimeters Hg, and water boils at 37°C (98.6°F). At this altitude atmospheric pressure is the same as the vapor pressure of water at the temperature of the human body. In other words, the fluids in the unprotected human body will boil. Thus, at extremely high altitude the very *milieu interieur* will boil away just from the heat of metabolism.

This physiological limit bears the name the *Armstrong line* after its discoverer, aviation medicine pioneer Major General Harry G. Armstrong (1899–1983), founder of Aero-Medical Laboratory at Wright Field in Dayton, Ohio, and later surgeon general of the U.S. Air Force. At the Armstrong line water can no longer exist in the liquid state at body temperature. As liquid water enters the vapor phase, the metabolic gases $O_2$ and $CO_2$ dissolved in the body fluids also come out of solution. This process, known as ebullism, is rapidly fatal because large bubbles of gas damage tissues and form in the blood, where they obstruct the circulation. The outer dimensions of the body may suddenly and dramatically increase, and unconsciousness ensues in a matter of seconds. Death inevitably follows within two or three minutes unless the pressure around the body is immediately brought back to normal. It is truly remarkable that undergoing a catastrophe of this magnitude, that is, the body's main constituent, water, undergoing a change in physical state, does not produce immediate death. Ebullism is a potential danger to the occupants of high-altitude balloons, aircraft, and spacecraft that depressurize at altitudes above the Armstrong line. Thus, aboard the space shuttle or the international space station the danger of losing cabin pressure is not just loss of oxygen, but ebullism as well.

The atmosphere at 62,800 feet above the Earth exerts a pressure (62.8 millibars) less than 7% of the normal pressure at sea level (1000 millibars). This pressure is six to ten times the atmospheric pressure on Mars (6 to 10 millibars), depending on Martian latitude and time of year. Thus, the body fluids of an unfortunate astronaut whose spacesuit depressurized on the surface of Mars would boil, and he or she would die of ebullism within minutes, even with adequate protection from cold and an artificial method for delivering oxygen to the tissues. Even if body temperature is allowed to fall by 10°C, unprotected humans cannot survive above 65,000 feet, much less on the surface of the Red Planet! If future engineers ever seriously consider providing a terrestrial atmosphere for Mars, their task will be to raise atmospheric pressure by at least twentyfold, or to an equivalent of Earth's upper troposphere. This pressure is needed to enable human beings to survive without wearing protective pressure suits. Inadequate atmospheric pressure is one of the most formidable problems that face future human endeavors to colonize planets with limited atmospheres.

## The Pressure Suit

Even before the invention of jet aircraft, the capabilities of the propeller-driven airplane could take it above 40,000 feet, and two physiological limits imposed by a rarified atmosphere, hypoxia and decompression sickness, were a reality for pilots. The final threshold at the Armstrong line could be crossed only by balloons or with the use of jet- or rocket-powered engines. Because the nature of these limits is determined by aspects of the physics of gases to which human adaptation is impossible, the problem of flight into the stratosphere and beyond without cabin pressure was solved by engineering a pressure suit. Pilots of today's high-altitude reconnaissance aircraft, such as the SR-71 Blackbird (Fig. 16.3), which operates at a service ceiling of 85,000 feet, routinely breathe 100% $O_2$ and are equipped with full pressure suits.

The inventor of the pressure suit was Russell Colley, an engineer for the B.F. Goodrich Tire Company in Akron, Ohio. In the 1930s Colley was asked by Wiley Post to make a suit for high-altitude flight. Colley's solution was to devise a suit of two layers—one for pressurization and one to keep the suit in a workable configuration. On September 5, 1934, Post, wearing a Colley suit, flew his airplane

190  THE BIOLOGY OF HUMAN SURVIVAL

**Figure 16.3.** The SR71 Blackbird is the highest flying jet aircraft in the world. It has a service ceiling of 85,000 feet, which allows it to fly above modern air defenses. The Blackbird has also served as a research platform for studies in high-altitude flight. The two crew members wear full pressure suits. (Source: National Aeronautics and Space Administration.)

to the record-breaking altitude of 42,000 feet. Russell Colley later created a segmented pressure suit, which featured pressurized pockets that greatly improved comfort for the aviator. Colley eventually worked for NASA, and his segmented suit concept was adopted by the agency's spacesuit program.

The pressure suit applies gas pressure to the surfaces of the body evenly, which prevents hypoxia, decompression sickness, and ebullism. The operating principle of the suit is that of Pascal's law, which states that the pressure surrounding a volume of fluid, such as the human body, is transmitted equally throughout all parts of the fluid. This means that the force of a gas between the body and the suit dictates the behavior of gas dissolved in the body fluids. The pressure and volume of the gas contained by a pressure suit must be kept relatively constant regardless of how high the aircraft flies. For this, the suit must have a dynamic control system that allows it to be maintained at a safe pressure. In space when the occupant is not connected to the spacecraft's life-support systems, the suit also must be entirely self-contained, or closed.

As Russell Colley recognized in the 1930s, in order to hold pressure the suit must consist of one or more internal layers of a gas-impermeable material. Another requirement is for an outer layer to prevent overdistension of the suit. These requirements mean that the suit tends to become rigid when inflated, like an automobile tire and inner tube, and this makes movement cumbersome.

The absolute pressure inside a modern pressure suit can be adjusted to the needs of the aviator or astronaut, but it is generally kept between 141 millimeters and 282 millimeters Hg. For example, the spacesuit used by NASA maintains an internal pressure of 4.3 psi, or approximately 0.29 ATA (222 millimeters Hg), equivalent to about 30,000 feet of altitude. Because air at a pressure of 222 millimeters Hg provides an ambient $PO_2$ of only 46 millimeters Hg, astronauts breathe pure $O_2$. This provides an inspired $O_2$ concentration greater than is that of air at sea level. Inside the space shuttle and space station, which fly at cabin atmospheres of 1 ATA (14.7 psi, or 760 millimeters Hg), astronauts preparing for extravehicular activity (EVA) also prebreathe $O_2$ for several hours before donning spacesuits. Meanwhile, the cabin pressure is reduced to 10.2 psi (527 millimeters Hg). This stepwise reduction in pressure and $O_2$ prebreathing eliminates $N_2$ from the body and helps prevent decompression sickness when the pressure in the suit is reduced to 222 millimeters Hg. Future NASA suits are being designed to operate at a pressure of 8.3 psi (0.56 ATA), which should prevent decompression sickness altogether.

These pressurized suits are very cumbersome and greatly restrict the mobility of the astronaut. This problem is manageable at zero gravity, but it would be very difficult to operate in such a suit on the surface of a planet like Mars. New suit designs that solve the internal pressure–external movement dilemma will be necessary for interplanetary exploration, but the solution cannot involve lowering suit pressure below about 130 millimeters Hg because of hypoxia.

The gas-impermeable inner layer of the pressure suit also solves a problem and creates another because it greatly limits the capacity for heat exchange. Therefore, the heat of metabolism is difficult to dissipate, and a heat load is imposed on the body. To keep the occupant cool, a large flow of gas is required to ventilate the suit. In an aircraft pressure suit airflow can be taken from the engines, but in a spacesuit operating in a near vacuum this method is unavailable. Instead, a recirculating system is used, and the air in the suit flows through a backpack, which removes $CO_2$, moisture, and heat and adds an appropriate amount of oxygen. Heat is also removed by a liquid-filled cooling garment worn next to the skin under the spacesuit. Liquid cooling is much more efficient than is air or gas cooling. Using these innovations in a complete package called an extravehicular mobility unit (EMU), astronauts are provided an integrated environmental protection, life support, and communications system that enables them to work safely outside the spacecraft (Fig. 16.4).

Spacesuits clearly prevent hypoxia, decompression sickness, and ebullism, but the problem of metabolic heat production means heat must be dissipated from the suit to the near vacuum of deep space, which presents another challenge. In Earth orbit the astronaut floats freely in space without an appreciable atmosphere. Lack of an atmosphere and lack of contact with a body of larger thermal mass means

**Figure 16.4.** Free-floating astronaut in low-earth orbit. The spacesuit contains complete life-support capability for short duration, including protection from radiation, insulation from heat and cold, adequate atmospheric pressure, an $O_2$ supply, and capacity for $CO_2$ absorption. Spacesuits are hybrid life-support systems, with many open features that reduce cost and improve reliability. For example, metabolic $CO_2$ is absorbed and later discarded, the $O_2$ necessary for the astronaut's activity is stored, not generated, in the suit, heat is dissipated to space by radiation, and wastewater (e.g., sweat and urine) is not recycled. (Source: National Aeronautics and Space Administration.)

no heat is lost from the suit by conduction, convection, or evaporation. Thus, cooling must be done by radiation. Unfortunately, the Sun, the largest body with which to exchange heat, radiates far more heat than does an astronaut's body. Thus, a spacesuit gains heat rapidly when an astronaut is exposed to direct sunlight.

A workable solution to this problem can be found by choosing the color of the suit. If the suit is black, the outer garment will heat to approximately 80°C from absorption of radiant heat from the Sun. If the suit is highly polished like a mirror and reflects the bulk of the Sun's rays, the problem becomes worse, not better, because the suit will not radiate heat to the environment. However, if the suit is white, which reflects visible light but retains its characteristics as a blackbody in the infrared region of the spectrum, a reasonable compromise can be achieved: the suit reflects visible and absorbs infrared wavelengths of light. Thus, the surface temperature of a pure white spacesuit in the vicinity of Earth stays at a manageable 55°C (131°F).

# 17

# G Whiz

The human body, like that of all terrestrial animals, evolved under the influence of Earth's gravity. In an evolutionary sense, physiological adaptation to gravity in vertebrate animals has been best exploited by birds. For humans and other ground-dwelling vertebrates, relatively modest deviations above or below normal gravitational field strength disturb physiological homeostasis. Life scientists do not yet agree on the principles that govern physiological effects in greater or lesser gravitational fields than that on Earth. What they do agree on is the importance of the hypothesis that a continuous relationship exists between the effects of extremes of gravity on the body.

## The Continuity Principle

The concept that hypergravity and microgravity are two extremes of a spectrum is known as the continuity principle. The issue of whether this principle reflects the true nature of the adaptive responses to high and low gravitational force is being thoroughly scrutinized because it may be critical to the success of long-duration space flights. The operating assumption is that the adverse effects of microgravity can be counteracted by acceleration, or hypergravity. Despite a great deal of scientific research, there is no consensus about the utility of the continuity principle, and like many general theories in biology, it seems to fit some types of responses but not others. The continuity principle holds interest for scientists concerned with the effects of terrestrial gravity (g) and for those concerned with the biology of inertial fields (G).

Loss of adaptation to gravity in space flight first became a practical reality on the prolonged missions of Skylab and the now defunct Russian space station MIR. Pictures of weakened astronauts being carried bodily off the Space Shuttle upon returning from weeks of near weightlessness in space once received great attention in the popular press, but equally impressive and important is the ability of humans and other animals to accommodate to sustained G by recruiting normal homeostatic responses. This chapter describes some of the problems of acceleration and high G forces, while the next deals with the effects of microgravity on the body.

## Gravity and Acceleration

Acceleration is useful not only in studying the physiological effects of Earth's gravity; it is the only practical approach to understanding the biology of hypergravity. When acceleration acts on the body, it creates an inertial force equal to the attractive force of gravity. This theory of equivalence was first put forward by the eclectic Austrian physicist Ernst Mach (1838–1916), whose name is more commonly associated with his work in supersonics. In practice, experiments on acceleration are conducted in ground-based centrifuges. Such studies can last just a few seconds or as long as several weeks to months. The most important studies in human physiology, however, have yet to be conducted and may take years to complete because development of critical effects of high-gravity environments may ultimately require very long periods of time.

The ability of the body to tolerate acceleration is known as G tolerance, which has three components: intensity, direction, and duration. Hence, it is generally placed in the context of the duration of an exposure to a force of a specified intensity and direction. High G forces in which duration of tolerance is very short, such as a few seconds, are considered separately in the next section. The duration of G tolerance also depends on biological factors, including the size and homeostatic capabilities of the animal species. Generally speaking, the duration of G tolerance is inversely related to body mass. Thus, very small mammals are able to compensate for several G indefinitely, while humans appear to have the capacity to tolerate only 1.5 G indefinitely. Laboratory rats can tolerate long-term exposures of 2 or 3 G because after a few days their bodies adapt to the sustained increase in inertial force. Chimpanzees, on the other hand, can tolerate 2 G for only a matter of a few hours.

Mammals adapt to chronic acceleration by changes in composition, muscle mass, work capacity, and food requirements. Because these changes require considerable time to develop, most of the information about them comes from animal studies. They will be discussed in more detail in the next chapter in the context of counteracting undesirable effects of microgravity. For now, it is sufficient to point out that the load-bearing components of the musculoskeletal system are most responsive to long-term changes in gravitational force, and this is true of both microgravity and hypergravity.

The basic principles of gravitational adaptation are demonstrated by the comparative physiology of terrestrial animals. For load-bearing land animals, muscle and skeletal mass increase about 5% for each kilogram of increase in body weight. Similarly, the musculoskeletal systems of small animals adapt to a 2G environment by increasing bone and muscle mass. The ratio of antigravity, or extensor, muscles to flexors increases, which improves exercise performance and imparts greater resistance to fatigue. In contrast, chronic exposure to near-zero-G environments decreases bone mass and extensor-to-flexor muscle ratio by an amount roughly equal to the increase found in the 2G environment.

In some animals body composition is modified rather dramatically in chronic hypergravity. Some of the change in composition has been attributed to the increased physical work required to maintain normal posture and locomotion. Generally, hypergravity decreases body mass despite increasing bone and extensor muscle mass. This occurs by loss of body fat. For example, in chickens, the percentage of body fat may fall by as much as tenfold during long-term exposure to 3G. This gravitational "defatting" appears to be related to a metabolic change that causes greater utilization of the simple sugar glucose and decreased conversion of glucose to fat. Gravitational defatting, however, is not the same for all species and has not been observed in primates.

## High-G Environments

In contrast to low-level acceleration, the dynamic high-G environments of high-performance aircraft and rockets are too stressful for the human body to withstand continuously. In aerial combat advantage goes to the most maneuverable aircraft, but this also imposes the greatest acceleration forces on the pilot. Modern high-performance aircraft easily exceed the capacity of pilots to tolerate the forces of acceleration, which culminate in G-induced loss of consciousness.

Human tolerance to high-G environments is too short, even with repetitive exposures, to expect much in the way of unique or sustained physiological adaptation. Rather, humans compensate for high-G forces primarily by physiological accommodation and trained behaviors. It is important to note that sustained G forces produce typical stress responses by increasing adrenal gland activity. These responses are useful in tolerance to subsequent acceleration and may have value in cross-adaptation, particularly with respect to hypoxia. If sustained G exposures are overly severe or frequent, physical exhaustion and eventual debility occur, characterized by damage to tissues. This pathology may also be involved in the genesis of certain kinds of adaptive responses.

In aviation and space flight temporary accelerations are classified according to the duration and direction in which they act on the body. For duration, forces that act for more than two seconds are considered long, while those of less than half a second are short. Short accelerations are usually encountered on impact with the ground or another solid object. Forces with durations between these times are considered intermediate accelerations. Physiological effects of intermediate- and long-duration accelerations are produced by deformation of tissues and organs and changes in the distribution of blood and other fluids in the body. Such accelerations may result from changes in either speed or direction, and they may be either linear or radial in nature.

The direction in which acceleration acts on the body is described by a system of coordinates defined in reference to the three main axes of the body. Thus, positive and negative accelerations, or +G and −G forces, are possible in the x, y, and z directions (Table 17.1). Much of the experimental work on acceleration has focused on the spinal, or z, axis of the body because it produces the most profound effects on the cardiovascular system. However, acceleration in the transverse, or x, axis, and, to a lesser extent, in the lateral, or y, axis, also creates important effects under certain circumstances.

The principles and nomenclature of acceleration also relate to deceleration from high velocity. Sudden deceleration, such as that produced in a high-speed impact with an immoveable object, transfers enormous force to the body. The rate of change in acceleration, or jolt, causes physical damage to the body because of the transfer of energy from the force of impact and uneven deceleration of different parts of the body that have different viscoelastic properties. When a vehicle carrying passengers such as an automobile or airplane, crashes the number of injuries and deaths relate primarily to the rate and duration of the deceleration period. The best predictor of injury and death of occupants is the instantaneous change in velocity ($\Delta V$).

Table 17.1. Three-Axis Nomenclature for Acceleration

| DIRECTION OF ACCELERATION (FORCE VECTOR) | AGARD* TERMINOLOGY | PHYSIOLOGICAL DESCRIPTION |
| --- | --- | --- |
| Forward (front to back) | $+G_x$ | Transverse G (A-P) |
| Backward (back to front) | $-G_x$ | Transverse G (P-A) |
| Right (right to left) | $+G_y$ | Left lateral G |
| Left (left to right) | $-G_y$ | Right lateral G |
| Headward (head to foot) | $+G_z$ | Positive G |
| Footward (foot to head) | $-G_z$ | Negative G |

*AGARD, Advisory Group for Aerospace Research and Development.

A good idea of the relationship between jolt and risk of death can be obtained from the analysis of automobile accident deaths, in which the probability of a driver being killed in a crash is related closely to the change in velocity during the crash (Fig. 17.1). The change in velocity is critical because velocity is the product of acceleration and time. Thus, the greater the velocity changes with time, the greater the force of deceleration on the body during the crash. Another glance at Figure 17.1 shows that it is highly unlikely for anyone to survive an automobile crash in which the instantaneous $\Delta V$ is greater than 100 miles per hour.

Experimentally, trained subjects with appropriate restraining harnesses have survived brief periods of extreme acceleration and deceleration. For example, on December 10, 1954, a U.S. Air Force flight surgeon, Colonel John P. Stapp (1910–1999), survived an acceleration of 19 G with deceleration of 40 G on a rocket-powered sled. Stapp's sled reached 632 miles per hour in 5 seconds and decelerated into a pool of water in 1.25 seconds. Rocket sled research of the 1950s conducted by the U.S. Air Force led to the development of modern safety harnesses, seats, and cockpit ejection procedures for supersonic aircraft.

## Limits of High-G Tolerance

The adverse effects of sustained high-G forces on the human body are related primarily to failure of the cardiovascular responses that compensate for the effects of acceleration. The propensity for cardiovascular embarrassment is most

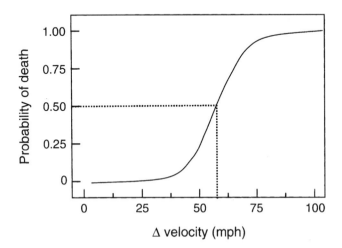

**Figure 17.1.** Impact–response curve for human crash fatalities. Probability of death (Y-axis) is plotted as a function of the instantaneous change in velocity during the crash (X-axis). The probability of death is 50% at an instantaneous velocity change of 60 miles per hour.

pronounced for G forces parallel to the z axis of the body. Other remarkable physiological effects of acceleration occur, such as decreases in lung volume and shifts in fluid balance, but cardiovascular factors are of paramount importance in G tolerance. Interestingly, the limit of G tolerance defined by critical failure of cardiovascular function is several times greater than is the terrestrial gravity to which the body has naturally adapted. This rather remarkable reserve capacity is the result of two facts. First, homeostasis of the normal circulation requires blood pressure to compensate for gravity effects during rapid changes in body position and for extreme changes in blood volume. Second, certain other stressors, such as hypoxia, may produce cross-acclimation useful for G tolerance.

Extreme changes in blood volume occur during dehydration and hemorrhaging, and being able to tolerate either condition clearly imparts a strong survival advantage to an individual. The same compensatory responses that maintain blood pressure during changes in blood volume are also used to maintain blood pressure during changes in posture. This response is known as orthostasis. Precisely the same set of cardiovascular responses is used to compensate for high-G forces during acceleration.

The application of high +G in the z-axis, the natural upright postural plane of a humans, produces serious cardiovascular system strain. This strain decreases the amount of blood return from body organs to the heart. This blood pools in the lower extremities because the veins of the legs are highly compliant. Thus, the critical function that fails under high G is cardiac output, which can no longer generate adequate blood pressure. The main consequence of low cardiac output is decreased blood flow to the eyes and brain, which manifests itself in gradual loss of peripheral vision, known as gray out. If acceleration is too severe or maintained for too long, complete loss of vision, or blackout, occurs, soon followed by loss of consciousness. These effects are graphed in Figure 17.2A.

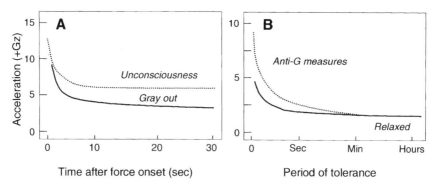

**Figure 17.2.** Effects of acceleration and anti-G measures on loss of visual function and consciousness. *A* indicates that symptoms of gray out and loss of consciousness occurs within seconds of exposure to G forces above 4 G. *B* illustrates the small advantage offered by using anti-G measures.

In humans G tolerance in the z axis is limited by the vertical height of the column of blood between the heart and the head. In other words it is inversely related to the minimum arterial blood pressure that must be generated by the heart × the density of the blood ÷ the vertical height between the heart and the head. The validity of this relationship has been confirmed in centrifuge experiments involving humans. The agreement of this prediction equation with experimental observations is remarkably close and suggests it is a very reasonable description of the physiological mechanisms.

Unprotected relaxed subjects normally tolerate an acceleration of only 3 to 4 G. At 4.5 $G_z$ the effects of acceleration on the normal hydrostatic forces in the circulation reduce arterial blood pressure at the brain to a value near to zero, and thus cerebral blood flow (CBF) is also close to zero. Because the brain depends on a continuous supply of $O_2$, the period of useful consciousness at 4.5 G is only about five seconds. At G forces of 3G to 4.5 G, the arterial blood vessels in the brain dilate by means of reflexes that are recruited to sustain the CBF. The veins in the upper neck also act as a powerful siphon to pull blood through the cerebral vessels. The siphon effect fortunately also drains the cerebrospinal fluid, causing the pressure inside the cranium to fall. This means the actual pressure available to perfuse the brain, the difference of arterial pressure − intracranial pressure (cerebral perfusion pressure), is maintained at 50 to 60 millimeters Hg. This value is near the normal value of 65 millimeters Hg. As acceleration approaches 5 G, however, the veins in the neck collapse completely, thereby breaking the siphon and arresting blood flow to the brain.

The changes in the lungs during acceleration are rather minor up to 5 G and generally do not limit the ability of a pilot to tolerate the exposure. The lower portions of the lungs, however, become progressively underventilated while remaining perfused with blood. Thus, some of the venous blood that enters the pulmonary artery is distributed to areas of the lungs that are not ventilated with fresh gas, and $PO_2$ in this blood does not increase normally as it passes through the pulmonary capillaries. When this blood returns to the left side of the heart, it dilutes the $O_2$ saturation of the hemoglobin in the entire left (arterial) side of the circulation. This creates a right to left shunt, which at 5 $G_z$ amounts to almost half the output of the heart. This shunt can lower the $O_2$ saturation of the arterial blood from its normal value of 98% to a dangerously low value of 85%.

Human tolerance to high-G forces is modest compared to that of some vertebrate animal species. For example, the peregrine falcon, which is capable of stooping at more than 300 kilometers per hour (190 miles per hour), may briefly sustain very high rates of acceleration. A simple calculation indicates that the falcon in a maximum dive remains conscious during accelerations of more than 7 G.

The point when G tolerance begins to fail can be overcome only by anti-G maneuvers. The most important anti-G techniques involve methods that maintain arterial blood pressure during high-acceleration maneuvers. Three anti-G techniques are widely used in high-performance aircraft: the anti-G straining maneuver, or AGSM; continuous positive pressure breathing, or CPPB; and anti-G suits. These measures can roughly double human tolerance to acute G forces (see Fig. 17.2B).

The circulatory effects of high acceleration became clear less than two decades after the Wright brothers' first flight, but the problem of loss of consciousness became significant only after the introduction of high performance airplanes in World War II. A practical anti-G suit to protect fighter pilots was invented in 1942 by Canadian scientist Dr. Wilbur R. Franks. The Mark III Franks Flying Suit was the first anti-G suit used in combat aviation and provided a substantial tactical air advantage to the Allies.

The AGSM has also been used to increase G tolerance in pilots since World War II. Because proper straining is learned in training, the AGSM is another good example of a behavioral adaptation. The benefit of learning the straining maneuver is also behavioral because it enhances the performance of the pilot by preserving vision and mental function. The performance advantage of straining, however, is achieved by a physiological mechanism.

Straining maneuvers produce an immediate increase in pressure in the thorax, which raises the blood pressure inside the heart by as much as twofold and may increase G tolerance by as much as 2 G. Using the maneuver with an anti-G suit, which forces blood from the legs into the chest, trained aircraft pilots can tolerate 6 G for two minutes, 8 G for one minute, and 9 G for thirty seconds. Such high-G exposures are very strenuous and lead to significant fatigue, which soon debilitates the aviator if the intensity, frequency, or duration of the maneuver is prolonged.

The application of CPPB to assist tolerance to $G_z$ is also helpful. However, measures for protection by CPPB at high G require special aircraft design considerations. The delivery of CPPB to the mouth increases the pressure inside the lungs and thorax, some of which is transmitted to the chambers of the heart and results in a direct increase in blood pressure. However, high intrathoracic pressure also impinges on the large central veins of the chest, the vena cava that return blood to the heart, and causes them to collapse. This decrease in venous return to the heart impairs the cardiac output. Thus, CPPB must be used with a system to support the venous return to the heart. The most effective way to accomplish this is by synchronized inflation of the anti-G suit, which is coordinated with the CPPB valve to avoid initiation of CPPB without first inflating the suit and forcing more blood into the vena cava.

The engineering performance of modern jet fighters allows them to exceed the physiological capabilities of a pilot even with state-of-the-art anti-G technology. Acceleration forces of 12 G are well within reach of these aircraft. In future high-performance aircraft designed to operate at such high-G forces, the position of the pilot's body will have to be altered to minimize $G_z$ force. An efficient way to do this is to place the pilot in a reclined seat, which would require that cockpit control systems be redesigned to operate from a recumbent position.

## Adaptation to Sustained G Forces

The mechanisms of homeostasis and adaptation to sustained G forces are unique, and they have important implications for prolonged human space flight. It is not surprising that chronic G-force tolerances are far below the acute tolerance limits, and the plot of G tolerance versus time is the familiar hyperbola. The human G-tolerance curve approaches its zero time asymptote at about 9 G, and its lifetime asymptote is about 1.5 G. Although this curve outwardly appears to be smooth, it is important to realize that human G tolerance, like other physiological strains, is limited by different physiological factors at different levels of G stress. Human volunteers have tolerated 1.5 G for seven days with no apparent ill effects. However, after just twenty-four hours at 2 G, evidence of significant fluid imbalance is detectable. At 3 G to 4 G fatigue is limiting, and above 4 G cardiovascular factors limit G tolerance.

What does a rather modest ability to tolerate sustained G have to say about the future of prolonged human space travel? The question is important if humans are ever to travel successfully outside the solar system. The answer, based on current physiological information, is sobering but not discouraging. To appreciate the problem, imagine a hypothetical spacecraft with a drive capable of accelerating it to a velocity of half the speed of light. The colossal logistical and relativistic problems of pushing or pulling along an adequate supply of fuel are well known, not to mention supplies of oxygen, water, and food, but these issues will be put aside for the moment.

The problem of acceleration can be illustrated as follows. The normal force of Earth gravity produces a constant acceleration of 9.8 meters per second squared, or 1 G. Thus, an acceleration of 1.5 G is 14.7 meters per second squared. The acceleration, A, is defined by velocity, that is, in meters per second ÷ time in seconds:

$$A = v/t$$

Because the velocity of light is 300,000 kilometers per second, or 300 million meters per second, it would require 10.2 million seconds at an acceleration of 14.7 meters per second squared to reach 150 million meters per second, or half the speed of light (assuming, for argument's sake, that the ship instantaneously achieves the desired acceleration). The final velocity would be reached after 118 days, or nearly four months, of constant acceleration. So far, the longest human experimental exposure to constant acceleration of 1.5 G has been only seven days.

Even a hypothetical spacecraft traveling at half the speed of light to the closest star that *might possibly* have a planetary system, the red dwarf Lalande 21185, at 4.2 light years away, would require eight and a half years to make the journey! A voyage to a nearby star that realistically may have a planetary system, such as

Epsilon Eridani, at 10 light years away, would require a journey of more than twenty years. As the spacecraft approached its destination, the occupants would also have to sustain negative acceleration for nearly four months.

A more physiological plan, to accelerate at only 1 G, would provide a "normal" gravitational field. In this case the ship would reach its final velocity in 174 days, and deceleration would take an equal amount of time (and power). Thus, the time spent on acceleration and deceleration alone would consume almost an entire year. On a 10 year voyage, however, this strategy would expose the astronauts to 9 years of weightlessness. Scientists must learn a great deal more about gravitational physiology before people can be subjected to such extraordinary journeys. This problem, however, would be a minor inconvenience compared to the total time, fuel, and other resources that would be required to traverse such astronomical distances.

# 18

# The Gravity of Microgravity

On April 12, 1961, 27-year-old Soviet cosmonaut Yuri Gagarin entered his tiny Vostok spacecraft atop the massive R-7 booster and rocketed into Earth orbit in his legendary three-orbit, 108-minute flight. Meanwhile the Western medical establishment was in the midst of a debate about whether human beings could tolerate extended periods of near weightlessness. The dramatic physiological effects of microgravity are multifaceted, but over the short term they are tolerated remarkably well by the human body. Prolonged weightlessness does not seem to impose a clearly identifiable limitation to survival, but it has been the source of concern in at least six areas of human biology: space sickness, cardiovascular function, lung function, immune function, loss of bone and muscle mass, and interactions with radiation.

A great deal of effort has been devoted to studying the effects and mechanisms of microgravity, but nearly half a century into the Space Age, debate continues about the mechanisms for some of them and about how important they will be in future missions of long duration (Fitts et al., 2000). To date the longest space mission has been the 437-day epic of Valeri Polyakov, the Russian physician–cosmonaut aboard the MIR station in 1994 and 1995. Indeed, as of 2002 only sixteen people had spent a total of more than one year in space (Fig. 18.1). However, such missions have provided evidence that certain effects of microgravity have more important implications than do others for long-duration spaceflight.

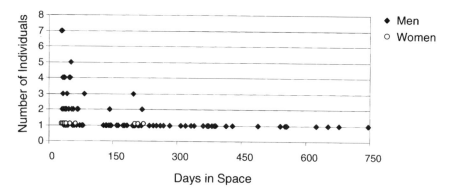

**Figure 18.1.** Human experience in spaceflight. Small diamonds indicate number of individuals and duration for all human spaceflights through 2001. Approximately three-quarters of all flights have lasted less than six months.

## Space Sickness

Space sickness is a form of motion sickness that has its origins in the peripheral vestibular system, which is required for balance and spatial orientation. The vestibular system consists of the otolith organs and the semicircular canals. The otolith organs relay information to the brain about gravity and linear acceleration that is coupled with input from the semicircular canals about acceleration in three dimensions to provide information about head movement and position. Microgravity appears to generate a mismatch in the information received by the brain from these two parts of the vestibular system when the head moves, which causes the spatial disorientation of motion sickness. The weightless condition forces the astronaut to learn to recalibrate the inputs from the peripheral vestibular system. Thus, space sickness is common but self-limited, appearing soon after entry into space and usually resolving within a week or ten days. Although the effects of space sickness can be dramatic, including severe nausea and vomiting, astronauts soon learn to accommodate, and it is not a cause of long-term disability in spaceflight.

## Intolerance of Upright Posture

Another dramatic effect of microgravity is intolerance of upright posture after returning to Earth from orbit. This orthostatic intolerance was identified very early in both the Russian and American space programs. Orthostatic intolerance can be so extreme that standing up severely compromises blood flow to the brain, and the individual suddenly loses consciousness. Because humans are relatively tall upright primates who have lived under the influence of gravity for thousands of years, people are well adapted to regional differences in the fluid, or hydrostatic pressure, inside the blood vessels—low pressure in the upper body and high in the lower body. The upright pressure difference between head and feet can be as much as the height of about a four-foot water column (90 millimeters Hg). When standing on the surface of the Earth, the neutral point of the circulation, or hydrostatic indifference level (HIL), occurs at the level of the diaphragm, just below the heart. At the HIL the vascular pressure remains constant whatever the orientation of the body with respect to the ground. However, in space the entire circulation becomes hydrostatically indifferent. In other words, there is no longer a hydrostatic pressure difference between head and feet in any position.

Scientists have used several ground-based strategies to attempt to understand the responses of the human circulation to hydrostatic indifference. These include parabolic flight, water immersion, and bed rest. All three strategies simulate microgravity reasonably well by minimizing hydrostatic pressure differences in the circulation, but they all have limitations. In parabolic flight the period of near weightlessness lasts only a few seconds, and immersion in water is impractical for more than a few hours. Only bed rest can be prolonged for days or weeks to simulate microgravity. Bed rest for three to four weeks in a 6° head down (tilt) position recapitulates many, but not all, of the cardiovascular, bone, and muscle changes of actual spaceflight. In small animals chronic suspension of the hindquarters in a harness has a similar effect. Although the information gleaned from ground-based experiments has proven valuable, there are discrepancies between the results of ground studies and data from actual spaceflights (Edgerton and Roy, 1996).

In flight astronauts show a number of circulatory adjustments to microgravity that can appropriately be called acclimation. These include cranial redistribution of fluids, reduced ECF and plasma volume, increased compliance of the leg veins, and decreased red blood cell mass. The outward manifestations of these effects are facial swelling, a decrease in calf dimensions (bird-legs), and, upon return to normal gravity, intolerance of upright position and lower capacity for physical exercise.

Loss of conditioning in space is a complex problem due in part to the decrease in plasma volume, which requires a greater increase in heart rate with progressive exercise than when plasma volume is normal. The effect is comparable but not identical to loss of plasma volume with immersion and dehydration. Thus, the hypothesis has been considered that the central fluid shift of microgravity could induce diuresis and natriuresis in the first few days in space, but recent in-flight experiments have found no evidence of these effects. In a few instances sodium retention has been reported with no fluid retention. The magnitude of the decline in plasma volume increases with the duration of flight and stabilizes at about 15% below normal after one or two months in space. The circulatory consequences of returning to Earth with low plasma volume are corrected partially by infusing intravenous fluid into the circulation before an astronaut's re-entry into the gravitational field.

## Loss of Bone Mass in Space

At the other end of the spectrum of microgravity's effects are losses in bone and muscle mass, which are insidious, progressive problems. Surprisingly little is known about the causes of these effects, but for prolonged missions in space the need to develop effective countermeasures is undeniable (Turner, 2000). This assumes, of course, that the astronauts intend to return to Earth or some similar planet to live out the rest of their natural lives. If the plan is to remain permanently in microgravity, then the negative consequences of the insidious loss of bone and muscle mass become moot.

To understand why bone loss is such an important problem in space, it is necessary to know something about normal bone metabolism. In young people the normal skeleton responds to increased or decreased mechanical forces by slowly increasing or decreasing calcium balance, thus increasing or decreasing bone mass. In older people there is an age-related decrease in bone mass that load bearing slows and unloading accelerates. In spaceflight load bearing is decreased, and bone mass gradually decreases. Although this change is commonly referred to as bone loss, it is more accurately termed *skeletal adaptation*. Skeletal adaptation is linked integrally to the Earth's gravitational force because the emergence of vertebrates out of the oceans onto land required the development of limbs strong enough to support the body's weight.

Gravity is essential to the development, growth, and proper maintenance of the skeleton in terrestrial animals, as is illustrated by the general relationship between the mass of a land animal and the size and strength of its weight-bearing bones. Larger animals must have larger and stronger bones because of the higher skeletal forces imposed by greater body mass and inertia. This principle of allometry was known by Galileo, who recognized that more and more of a material of constant mechanical properties is needed to support greater weight until the structure eventually collapses under its own weight.

Bone size correlates roughly with body mass, but the strength and geometry of the skeleton appear to have evolved more particularly under the influence of maximum dynamic strain on the bones so as to avoid collapse, such as during high-speed maneuvers. Indeed, when body scale is taken into account, the bones of larger animals are not so strong as those of smaller animals, in part because muscle power per kilogram of body mass is greater in smaller animal. Thus, the biomechanics of bone and muscle are coupled in a complicated way not only to gravitational forces but also to peak mechanical stress on the bone. This coupling is essential in order for musculoskeletal performance to remain safely within the limits of the strength tolerance of bone and ligament.

The relationship between bone strength and dynamic strain indicates that bone responds at the cellular level to changes in external mechanical forces. Although this principle has been confirmed many times in the laboratory, the transducer mechanism(s) whereby a change in force brings into play an adaptive response in the bone is still a mystery. All three major types of bone cells, the bone-forming osteoblast, the bone-absorbing osteoclast, and the osteocyte encased in its mineral matrix, respond to mechanical stresses by changing their activities. The magnitude of the physical stresses on these cells, combined with the influences of calcium-regulating hormones of the endocrine system, determine the extent of mineral gain or loss in different regions of the skeletal system. Thus, the primary load on the axial skeleton and support limbs is generated by locomotion, and bone homeostasis is maintained normally under the ever-present force of gravity.

Complete growth and maturation of the skeletal system is achieved in early adulthood. Genetic factors are responsible for 60% to 80% of peak skeletal mass and bone density, but nutrition and lifestyle play important roles in bone growth. In adults maintenance of skeletal strength and repair of microscopic skeletal damage are accomplished by bone remodeling. This process is regulated by vitamin D; circulating hormones, including estrogens, androgens, and parathyroid hormone (PTH); and local growth factors. Other important influences include nutrition, particularly dietary calcium, and physical exercise. The process of remodeling replaces older or damaged bone with a similar amount of new bone. This involves two interrelated processes: resorption of old bone and formation of new bone. Thus, remodeling maintains a constant skeletal mass throughout most of adulthood. Sometime after age 35, however, the natural equilibrium between bone resorption and formation becomes altered, and the rate of resorption exceeds that of formation. This imbalance varies at different skeletal sites and is exaggerated in women after menopause. Loss of bone density from age-related changes and from medical factors that accelerate this process is called osteopenia or osteoporosis. These terms also accurately describe the long-term skeletal changes of microgravity.

The responses of the skeletal system to microgravity are similar to its responses to immobilization, which in elderly people greatly exaggerates the progression of osteoporosis. As every child who has ever broken a leg knows, the bone is slow to heal, and the circumference of the limb after the cast is removed is smaller than that of the other. These effects are the dramatic consequences of immobilization. Indeed, the unloading in microgravity is greater even than that of immobilization, and the long bones, which bear the greatest loads on Earth, experience the greatest effects of unloading in space. Less stress on the skeleton from load bearing means that less bone mass is needed to maintain skeletal architecture.

In the past the amount of bone mass loss in microgravity has been determined by measuring the loss of calcium from the body. More recently, direct measurements of bone mineral density (BMD) by imaging techniques such as QCT (quantitative computerized tomography) and DXR (dual energy X-ray absorptiometry) have become possible. The BMD determines most (about 70%) of the variation in bone strength. Hence, bone density measurements are useful for predicting the strength of bone and risk of fractures, particularly in elderly people. For example, in postmenopausal women bone density declines about 1% per year after age 65, which correlates with an increasing risk of hip fracture.

The normal adult body contains approximately 1.25 kilogram (2.75 pounds) of calcium, of which approximately 5 grams is in a freely exchangeable pool. During spaceflight body calcium balance, which is usually neutral, becomes negative. In other words, the body in microgravity loses a small amount of calcium from the exchangeable pool each day. A 1974 report on the eighty-four-day Skylab mission found an average loss of calcium of about 140 milligrams per day in three astronauts who used regular exercise countermeasures. This amounts to only about 1% of the total body calcium, but if this rate of bone loss were to persist throughout a two-year mission to Mars, it would amount to an 8% loss in body calcium and, on average, a similar loss in BMD. To place this into perspective, it has been found that a 4% to 5% decrease in bone density is associated with a roughly 15% lifetime risk of hip fracture in postmenopausal women over age 50. Thus, an astronaut returning from such a long mission would likely have brittle bones susceptible to stress fractures.

An equally disconcerting problem with loss of BMD in space is that it is not uniform but occurs primarily in the weight-bearing bones. For example, bones in the skull and the arm (radius) show no loss, while the lumbar spine, pelvis, and heel (calcaneus) show greater loss than would be predicted from the total excretion of calcium. Thus, it has become clear that effective countermeasures are needed to prevent weakening of the weight-bearing parts of the skeletal system in long-duration space missions. To make matters worse, there is great variability in bone loss among individuals in space, which makes it difficult to predict the magnitude of bone loss or the time necessary for recovery. The rate and extent of recovery are real concerns because in at least one study bone density had not returned to preflight levels five years after the mission.

Although microgravity alters calcium balance and increases bone loss, the cellular mechanisms are unknown. In-flight studies in animals and cells in culture have not fared much better than have human studies in shedding light on the problem. It is generally agreed, however, that skeletal growth is impaired in developing animals in space. In human adults the situation is not at all clear. Measurements of biochemical markers of bone formation and bone resorption have given conflicting data. Circulating markers of bone resorption usually increase, while those for bone formation may increase, remain the same, or decrease. Furthermore, no direct measurements exist of the strength and quality of new bone formed by humans in space. As a result, it is still not known if bone loss in microgravity is caused by failure of bone remodeling, acceleration of bone remodeling, or some other process or combination of processes. These conceptual gaps make it difficult to devise effective countermeasures or pharmacological therapies for the problem. In order to make a breakthrough in this area, a great deal more research will be needed.

## Loss of Muscle Mass in Space

In addition to the problems of bone loss in microgravity, weightlessness has been associated with a decrease in the mass (atrophy) of skeletal muscles (Convertino, 1996). The cellular mechanisms of this muscle atrophy are not well understood, but at least two important problems are recognized. First, spaceflight may cause muscle atrophy indirectly by altering circulating hormonal factors, such as growth hormone, stress hormones, and anabolic steroids. In addition, microgravity may have direct effects on the muscle fibers themselves. The development of atrophy reduces muscle functional capacity and increases muscle fatigue, particularly in the antigravity muscles of the back and lower limbs. In the first few days in microgravity, atrophy of extensor muscles is greater than in flexors, but by six months the degree of extensor and flexor atrophy is approximately the same. Some calf muscles, such as the soleus, have shown as much as a 20% decline in size after six months in space accompanied by substantial decreases in peak force and power. In addition, as with bone loss, substantial individual variability occurs in the extent of muscle atrophy during spaceflight.

In examining the types of muscle fiber atrophy produced in spaceflight, humans, unlike some other species, show roughly equal atrophy of the two major fiber types, type I and type II. Muscle contraction consumes energy in the form of adenosine triphosphate (ATP) generated by glucose (glycolysis) alone or by both glycolysis and respiration in mitochondria. The oldest and simplest classification of muscle fibers is based on their metabolic enzyme composition and the initial velocity of contraction, type I, or slow-twitch, and type II, or fast-twitch. Type I fibers are red because they contain an abundance of the heme pigment, myoglobin, to facilitate the diffusion of oxygen from capillary to mitochondrion for aerobic respiration. Type II, or white fibers, contain little or no myoglobin and rely primarily on glycolysis for ATP production. Most muscles are mixtures of these red and white fiber types arranged in a pattern suitable for the type of work they perform. Muscles that contract with great force over a short period consist of mostly white fibers, and those that perform sustained work contain mostly red fibers. The red fiber generates ATP using respiration, which is roughly a dozen times more efficient than is glycolysis and is readily sustained over a long period.

The muscle cell (myocyte) contains contractile elements, or myofibrils, composed of bundles of actin and myosin protein filaments arranged in a regular lattice of six actin filaments that surround each myosin filament. The thick myosin filaments form cross-bridges with the thinner actin filaments during contraction. This allows actin and myosin filaments to slide along each other and shorten the fiber. These cross-bridges form and break down as the muscle contracts and relaxes. Regular cycling of actin and myosin cross-bridges is necessary to maintain muscle cell homeostasis and muscle bulk, tone, and strength. Thus, nutritional and hormonal influences aside, muscle homeostasis is maintained primarily by two factors: voluntary work and input from the neuromuscular systems that sense and maintain position and balance of the extremities (proprioception). There is increasing evidence that both factors are affected by weightlessness.

Muscle atrophy in prolonged microgravity is caused by structural and biochemical changes in individual muscle cells that appear to result from muscle unloading in flight and not from muscle cell damage from reinstitution of load bearing after returning from space. After a short period of weightlessness, the rate of protein synthesis in the muscle cell declines. However, contractile proteins are lost faster than are other proteins, and thin actin filaments are lost faster than are thick myosin filaments. Losses in mitochondrial proteins occur to a lesser degree, and in some small animals the enzymes of glycolysis actually increase.

The loss of contractile proteins in spaceflight is associated with a decrease in peak muscle force, although the continuing loss of thin actin filaments may reflect natural adaptation that allows the initial velocity of muscle contraction to increase. The geometrical change increases the distance between the thick and thin filaments, which may permit muscle cross-bridges to detach sooner. In theory, this could increase velocity by reducing intracellular drag, or friction, as the fiber contracts. In addition, some muscle fibers increase their expression of the myosin "fast-type" isoform during spaceflight.

Molecular biology studies of skeletal muscles in microgravity indicate that the decline in contractile protein synthesis is due to decreases in both gene expression and translation of messenger RNA into protein. Beyond this, little is known about how muscle atrophy occurs in space. A fascinating study published in 1999 by investigators at Brown University showed that isolated skeletal muscle cells are directly affected by spaceflight. The scientists engineered avian skeletal muscle cells into artificial muscles and flew them in perfusion "bioreactors" for ten days aboard the Space Shuttle (Space Transportation System, STS). They observed muscle fiber atrophy as a result of a decrease in the rate of protein synthesis without increased protein degradation. Returning to Earth stimulated synthesis of muscle proteins relative to similar cells on the ground. This study gave the first hint that skeletal muscle fibers respond directly to spaceflight, presumably because of the effects of microgravity.

The rapid and pronounced skeletal muscle atrophy in microgravity clearly degrades human performance in space. A more exact understanding of the cellular and molecular mechanisms of this wasting is needed in order to devise better long-term countermeasures. To date, widely employed countermeasures that require one- to two-hour daily periods of aerobic exercise on stationary bicycles or treadmills have helped prevent loss of physical endurance but have had limited effect on muscle atrophy. These protocols will be insufficient for interplanetary spaceflight. In animal and some human studies, more effective countermeasures have included interspersed high-resistance isotonic and isometric exercises for short periods throughout the day. The optimal duration and frequency for such exercise periods has not been determined, nor is there much published information on alternative countermeasures, singly or collectively, such as electrical stimulation of muscles, nutritional approaches, or hormonal therapy.

A number of scientists have strongly recommended instituting gravitational countermeasures in space to more closely emulate conditions on Earth. Providing artificial gravity by rotating the spacecraft or using an on-board centrifuge has much to recommend it because of the potential to avoid both the known effects of microgravity as well as those that have not yet been discovered. In Chapter 20 some potentially important physiological differences among short-term spaceflight (less than one year), extended spaceflight, and multigenerational space voyages will be examined and compared.

# 19

## Weapons of Mass Destruction

The reason to cover weapons of mass destruction in a book about extreme environments is because they have the potential to forever alter life on earth. The inconceivable destructive power of modern thermonuclear devices is antithetical to civilization, which requires a sophisticated infrastructure. The deaths from detonation of even one nuclear warhead over a population center would dwarf the September 11, 2001, World Trade Center disaster.

The exploitation of fuel-loaded jets to attack the World Trade Center produced an unusual situation by trapping the victims inside the twin towers, whose collapse vaporized them. Usually, immediate deaths from massive explosions represent a fraction of the total, particularly if toxic gases or radioactivity are involved. The grim task is to separate those waiting to die from those who can recover with medical care. This principle of triage is similar for all types of weapons of mass destruction, whether they involve radioactivity, biological agents, or nerve gas. However, the problems of scale and loss of medical and transportation infrastructure to manage such disasters are seriously underestimated by triage strategies.

This chapter briefly summarizes some of the major biological and human problems of nuclear, biological, and chemical (NBC) warfare. It will become clear that the greatest threat to the survival of humankind remains thermonuclear weapons because biological and chemical agents cannot yet be fashioned into weapons of comparable scale. If not already apparent, it will also become clear that the idea of a "preparedness plan" for a full-scale thermonuclear exchange is irrational. In actuality, the detonation of even a single modern thermonuclear device in a heavily populated urban area would produce more casualties than could be handled by the *entire* health-care system of the United States. Solitary nuclear devices are likely to be used now more than ever because of the worldwide emergence of terrorism as a means of challenging socioeconomic and technological preeminence. However, before providing specific information about the effects of nuclear weapons, I will set the stage with an overview of the baleful doctrine of biological and chemical warfare.

## Biological and Chemical Warfare Agents

In the United States the public health aspects of transmissible diseases are the purview of the Centers for Disease Control (CDC). Historically, the CDC has maintained public health resources to identify and track emerging infections and provided reliable information on treatment to health professionals, but the agency also tracks potential biological and chemical warfare agents. After letters containing anthrax appeared in the U.S. Postal Service in 2001, the CDC was charged with expanding its monitoring and public education programs.

The official position of the CDC on biological and chemical warfare was stated in the *Morbidity and Mortality Weekly Report* of April 21, 2000: "The public health infrastructure must be prepared to prevent illness and injury that would result from biological and chemical terrorism, especially a covert terrorist attack. As with emerging infectious diseases, early detection and control of biological and chemical attacks depends on a strong and flexible public health system at the local, state and federal levels." This position statement is technically and politically correct but too ambiguous for planning responses or practical implementation. Most disaster plans are designed to deliver water, food, medicines, and blood products to a few thousand casualties of an earthquake or other natural disaster. They are not capable of containing or controlling the effects of the release of a huge cloud of "weaponized" anthrax spores over Manhattan, which could kill half a million people and expose millions more to a deadly infection.

Biological and chemical weapons have a long and checkered history. The use of biotoxins dates to the sixth century B.C., when the Assyrians poisoned enemy wells with rye ergot. Bacterial diseases such as plague, glanders, and anthrax and viral diseases such as smallpox have been used as crude biological weapons for at least 700 years by exposing enemies to corpses or contaminated materials. Chemical weapons arrived with the industrial revolution and achieved their greatest notoriety in World War I, when chlorine, phosgene, and mustard gas were used against troops in the trenches.

As medical science has advanced, numerous bacteria, viruses, fungi, rickettsiae, and toxins have been discussed as potential biological warfare agents. These agents have been considered a threat mainly against small groups rather than as weapons of mass destruction because they are difficult to fashion and deploy on a large scale. Using modern technology, however, certain microorganisms can serve as weapons of mass destruction. They remain a potential threat despite formal political attempts to prevent their development and dissemination. Even the United States had a biological weapons research program from 1943 until 1969, when an executive order of former president Richard Nixon ended it.

In 1972 the United States and many other countries signed the Convention on the Prohibition of the Development, Production and Stockpiling of Bacteriological and Toxin Weapons and on Their Destruction, often called the Biological Weapons Convention. The treaty forbade stockpiling of biological agents for military purposes and ended research on offensive biological agents. In 1971 and 1972 the biological agents and munitions left over from the defunct U.S. program were destroyed, including microbes that cause anthrax, tularemia, and brucellosis; viruses for Q fever and Venezuelan equine encephalitis; and supplies of staphylococcus and botulism toxins.

Despite the 1972 accord, biological warfare research continued in countries other than the United States, including the Soviet Union and Iraq, both of whom had signed the convention. Biological weapons have actually been used since 1972. There were "yellow rain" incidents in Southeast Asia in the 1970s involving T2-mycotoxin, the use of ricin for an assassination in London in 1978, an accidental release of anthrax aerosol into the air over Sverdlovsk in 1979, which killed sixty-six people, and the intentional spreading of anthrax spores through the mail in the United States in 2001. Since the dissolution of the Soviet Union and dispersal of its weapons, there has been increased concern over the possible terrorist use of biological agents to threaten military and civilian populations. Indeed, several extremist groups are actively trying to obtain supplies of microorganisms suitable for fashioning biological weapons (Alibek and Handelman, 1999).

The frightening potential of biological weapons was pointed out by the World Health Organization (WHO) as early as 1970 (Health Aspects of Chemical and Biological Weapons, WHO, 1970). WHO estimated that 23 pounds (50 kilograms) of aerosolized anthrax spores dispensed 1 mile (2 kilometers) upwind of a population center of 500,000 unprotected people in ideal weather would travel more than 10 miles (20 kilometers) and kill or disable as many as 125,000 people in the path of the cloud. If tularemia were dispensed, the number of dead and incapacitated was estimated to be roughly the same. In addition, a major urban attack could disable the health-care system with overwhelming numbers of people in need of emergency and critical care, specialized medications, and vaccines.

The potential of biological agents for use as weapons varies considerably. The most dangerous agents have been placed into three categories by the CDC (Table 19.1). High-priority agents (category A) include organisms of risk to national security because they are easily disseminated or transmitted person-to-person, cause high mortality, may cause public panic and social disruption, and require special public preparedness. Category B agents are lower priority because they are not so easy to disseminate and would cause moderate or low morbidity and mortality, although they would require specific enhancement of CDC diagnostics and disease surveillance. Category C includes emerging pathogens that could be engineered for mass dissemination in the future if they became available or easier to produce and disseminate or if they have a likelihood for high morbidity and mortality.

Classification of biological agents by perceived threat is unsatisfactory because it is based on yesterday's intelligence. Specific molecular virulence factors are associated with a high death rate and can be engineered into different organisms. Factors that define the true risk in terms of human survival include infectivity, contagion, and lack of effective therapeutic or prophylactic measures such as antibiotics and vaccines.

Serious infectious diseases either kill the host or the host survives thanks to innate immunity or treatment, or both. Observations in nature and in the laboratory indicate that a small number of healthy survivors of most species will soon repopulate an area decimated by disease. This is because natural infectious diseases, with rare exceptions (such as mumps, which occasionally causes male sterility), do not affect the long-term ability of a host to reproduce. To the life cycle of microorganisms, this makes evolutionary sense: destroying the host population also destroys the infectious agent. Major epidemics, however, may create population bottlenecks that decrease genetic diversity in the new population, which will arise from a few survivors.

Toxins are natural substances produced by microbes, plants, or animals that are poisonous to other forms of life. They have been distinguished traditionally from chemical agents such as sarin and mustard gas because they are not manmade, but this distinction is no longer meaningful because almost any toxin can be synthesized today in a laboratory. Mustard gas was used in World Wars I and II to incapacitate soldiers, and its use was reported in the Iran–Iraq war of the 1980s. The Iraqi army also used mustard gas and nerve gas to kill 5000 thousand Kurds in a civil uprising in 1988. Sarin, another highly toxic nerve gas, made worldwide headlines when members of a Japanese cult released it in a Tokyo subway in 1998, killing 12 people. Gases tend to be less lethal than are biological toxins at the same concentrations (see Table 19.2), but most toxins are not naturally volatile.

The potential of many toxins to be used as weapons is limited by the inherently low toxicity of aerosols (too much toxin is required to make an effective aerosol). However, dispersal agents designed for appropriately sized particles can allow

**Table 19.1.** CDC Categories of Biological Warfare Agents

| CATEGORY A | CATEGORY B | CATEGORY C |
| --- | --- | --- |
| *Bacillus anthracis* (anthrax) | *Coxiella burnetti* (Q fever) | Nipah virus |
| *Clostridium botulinum* (botulism toxin) | *Brucella* species (brucellosis) | Hanta viruses |
| *Yersinia pestis* (plague) | *Burkholderia mallei* (glanders) | Hemorrhagic fever viruses |
| *Variola major* (smallpox) | *Ricinus communis* (ricin toxin) | Tick-borne encephalitis viruses |
| *Francisella tularensis* (tularemia) | *Clostridium perfringens* (α toxin) | Yellow fever |
| Viral hemorrhagic fever | *Staphylococcus enterotoxin B* | Drug-resistant tuberculosis |

**Table 19.2.** Lethality of Common Toxins and Chemical Poisons

| AGENT | $LD_{50}$ (mg/kg*) | MOLECULAR WEIGHT (DALTONS) | SOURCE |
|---|---|---|---|
| Botulism toxin | 0.001 | 150,000 | Bacterium |
| Shiga toxin | 0.002 | 55,000 | Bacterium |
| Abrin | 0.04 | 65,000 | Rosary pea |
| Maitotoxin | 0.10 | 3,400 | Dinoflagellate |
| Palytoxin | 0.15 | 2,700 | Marine coral |
| Ciguaratoxin | 0.40 | 1,000 | Dinoflagellate |
| Textilotoxin | 0.60 | 80,000 | Elapid snake |
| Batrachotoxin | 2.0 | 539 | Poison arrow frog |
| Ricin | 3.0 | 64,000 | Castor bean |
| α-Conotoxin | 5.0 | 1,500 | Cone snail |
| Tetrodotoxin | 8.0 | 319 | Puffer fish |
| α-Tityustoxin | 9.0 | 8,000 | Scorpion |
| Hydrogen cyanide | 10 | 27 | Chemical agent |
| VX | 15 | 267 | Chemical agent |
| Sarin | 100 | 140 | Chemical agent |

*Values are for laboratory mice.

microbes to become highly lethal weapons. The prototype aerosol weapon of this type is anthrax, for which it has been estimated that the inhalation of as few as ten spores can be fatal. After inhalation the spores germinate in the mediastinal lymph nodes, and the bacteria multiply and produce a toxin called lethal factor. Lethal factor and two other toxins are responsible for the high mortality of inhalation anthrax.

Although the threat to the survival of an individual from exposure to a particular biological agent may be high, the primary risk for the extinction of all humankind comes from agents that kill everyone shortly after exposure or that permanently destroy germ cells. It is also possible to breed or genetically engineer microbial hybrids, or chimeras, that express new or strange virulence factors. New (or new forms of old) microbes, such as natural mutations or genetically engineered variants, pose the most important threat to world health. The results of manipulation cannot always be predicted or controlled. Hence, apart from their inherent malevolence, new biological warfare agents encompass the theoretical possibility of creating an ultimate pathogen inadvertently. The only option for survival under such circumstances is to quarantine the exposed individuals to contain the spread of the disease.

## Thermonuclear Weapons

The effects of detonating atomic bombs over two Japanese cities in World War II were so horrendous that it has served as the basis for détente for half a century. These bombs killed at least 103,000 and perhaps as many as 200,000 civilians.

Although nuclear testing continued until the 1963 Test Ban Treaty, the Cold War between the United States and the Soviet Union settled into a period of nuclear deterrence through mutual assured destruction (MAD). Since the fall of the Soviet Union, a massive nuclear exchange by superpowers seems less likely, but the detonation by extremists of a crude atomic device in a heavily populated area seems more likely. The three heads of the thermonuclear hydra, the massive blast, thermal fireball, and ionizing radiation, could kill millions and expose millions of others to deadly radioactivity. Also, atomic (fission) and crude thermonuclear (fission and fusion) blasts disperse large amounts of radioactive fallout that could expose many more people to radiation for decades to come.

The presence of radiation has always been a fact of life on Earth. People are bombarded constantly with radioactivity from the cosmos, from the planet itself, and by inhaling and swallowing radioactive materials. Radiation is received from radioactive potassium in foods. The lungs are exposed to radiation in air, which contains small amounts of radioactive radon. This is natural background radiation, and life has adapted to it over millions of years. Exposure to natural radiation is increased by both flying and mining and decreased in submarines, even nuclear ones. Commercial jet air travel exposes the body to an average radiation dose from space of 100 times the dose one receives on the surface. Nevertheless, natural radiation is inconsequential compared to that released by the detonation of nuclear weapons. Cosmic, planetary, and internal sources of radiation each contribute about one-quarter of the body's natural exposure. Radiation generated by human activities, including X-ray equipment, radioactive medicines, and atomic weapons testing, makes up the difference.

In 1952 the United States exploded the world's first fusion device (hydrogen bomb), which had 1000 times the energy yield of the Hiroshima fission bomb. Soon thereafter the Soviet Union detonated its first hydrogen bomb. The nuclear arms race was joined by Britain (1952), France (1960), and China (1964), all of which argued that they must have nuclear weapons if other nations did. Currently, more than forty nations have nuclear weapons.

In 1959 Physicians for Social Responsibility (PSR) analyzed a hypothetical attack on the United States to point out the consequences of a thermonuclear explosion in a populated urban area. PSR estimated that for the Boston area alone a single blast combined with radiation exposure would kill 1.3 million people and injure 1.3 million more. Widespread destruction of health care facilities would doom a million injured people. Since 1960 the population of metropolitan Boston has grown to nearly 6 million people. Although the urban population has actually declined, the suburban population has doubled. The obvious message for today remains the same: the number of immediate deaths from a nuclear explosion would be similar, but those exposed to deadly levels of radioactivity could be as high as several million.

The number of health-care professionals and hospital beds available to treat the casualties of a thermonuclear detonation is very limited. In the United States there are approximately 750,000 physicians, 2.1 million nurses, and 850,000 hospital beds nationwide. It would not be possible for this overburdened system to absorb 3 million or 4 million additional severely injured patients in a few days. On this basis, civil defense and disaster planning for a terrorist detonation of just one urban nuclear bomb could be viewed as futile. The only reasonable alternative is to work to prevent such a disaster.

In 1984 PSR and the International Physicians for the Prevention of Nuclear War published "The Counterfeit Ark," which refuted the plan of the Federal Emergency Management Agency to respond to thermonuclear war. They pointed out that "to accept the survival of 80% of the U.S. population as a reasonable policy goal is also to accept as reasonable the deaths of 45 million people." In the 1980s scientists also warned that a thermonuclear exchange between two superpowers might cause a nuclear winter that could threaten the extinction of the human species. More circumspect calculations have predicted that the prolonged climatic temperature drop would disrupt agriculture and lead to large pockets of starvation and disease. Worldwide interruption of agriculture, manufacturing, health care, and transportation could kill half the world's population. Such visions of the world make the strongest possible case for nuclear disarmament.

## Types of Radiation

Different types of radiation have different effects on living organisms. These effects can be predicted by measuring the radiation exposure using dosimetry. The standard unit of measurement, the roentgen (R), was originally defined using the effects of radiation in an ionization chamber. R is an amount of gamma rays or X rays necessary to ionize a specific volume of air under standardized conditions. For biological exposures the dose has been expressed traditionally in units of the radiation-absorbed dose, or rad, the energy deposited in a tissue (100 ergs per gram of tissue). Because more radiation is absorbed as it passes deeper into tissue, the rad represents the net deposition of energy in a three-dimensional volume of tissue.

The rad has been replaced by a Système Internationale (SI) unit, the gray (Gy), that is equivalent to 100 rads. To compare the effects of various types of radiation on humans, a unit of dose equivalence is used, known as the rem (roentgen equivalent in man). The SI unit, the sievert (Sv), is 100 rem. The rem reflects the exposure dose multiplied by a biological effect factor for the type of ionizing radiation. An instantaneous total-body dose of 10 sieverts, or 1000 rems, is lethal, while the average background radiation dose per year is below 100 millirems (mrem), or 1/10,000 a lethal dose.

When tissues absorb energy from radiation, their molecules undergo both excitation and ionization. Excitation raises one or more electrons in an atom or molecule to a higher energy state without ejecting the electron(s). Ionization ejects one or more electrons from the atom. Ionizing radiation is classified as either electromagnetic or particle radiation. X-rays and gamma rays are electromagnetic radiation from different sources. X-rays are produced when electrons strike a target and emit energy, and gamma rays are produced by the decay of radioactive isotopes. Gamma rays and X-rays, as "packets," or quanta, of energy and photons, which lack mass or charge, tend to travel in straight lines. Photon energy occurs in quanta of multiples of h, or Planck's constant.

Particle radiation, including electrons (beta particles), protons, neutrons, and alpha particles, have mass and charge (except neutrons). Charged particles can be accelerated in electrical fields, and beta particles, which are negatively charged, are small enough to be accelerated to nearly the speed of light. However, they all decelerate rapidly in tissue and therefore do not penetrate deeply. Protons are positively charged and have a mass of 1 (2000 times that of an electron). When a proton enters a tissue, its energy is given up abruptly, which tends to cause a region of enhanced ionization called the Bragg peak. This means that protons exert their effects in a relatively focused area. Like protons, neutrons have an atomic mass of 1, but, being neutral, they cannot be accelerated by electrical fields. Neutrons are emitted as fission products of heavy radioactive elements or are produced in colliders. Alpha particles are helium nuclei (two protons and two neutrons) of sufficient mass and charge that they do not penetrate into matter unless they have huge energies. A sheet of paper will block most alpha particles.

## Biological Effects of Radiation

Ionizing radiation damages tissues in several ways. These include pair production, the Compton effect, and the photoelectric effect. The photoelectric effect occurs at low energies when an incident photon interacts with an electron in an outer electron shell of an atom. If the photon is more energetic than is the binding energy of the electron, the electron is driven off with a kinetic energy equal to the energy of the incident photon minus its binding energy. The Compton effect occurs at higher energies when an incident photon interacts with an electron, part of its energy is transferred to the electron, and the rest continues on as a less energetic photon. Pair production occurs when high-energy photons are absorbed and produce both a positron and an electron in the absorbing material. A positron has the same mass as does an electron with a positive instead of a negative charge. Positrons rapidly combine with electrons and annihilate both particles, which emits two photons in opposite directions.

The implication of these effects is that different types of ionizing radiation of the same dose produce different biological effects. In addition, certain particles, such as neutrons, produce a greater biological effect than do X-rays. Biological effects produced by a specific radiation dose can be compared using the relative biologic effectiveness (RBE), which is a quality factor referenced to a standard 250-kilovolt photon. The RBE for X and gamma rays is one. Generally, a greater RBE indicates a greater biological effect, and values are highest for highly ionizing radiation, such as neutrons, which have an RBE of 10 to 20.

RBE depends on radiation dose and dose rate, the type of tissue exposed, and the linear energy transfer (LET), or the amount of ionization that occurs along the length of the radiation path (expressed in kilovolts per micrometer). High- and low-LET radiation produce biologically different effects that are particularly noticeable when different amounts of oxygen are present in cells. Well-oxygenated and poorly oxygenated (hypoxic) cells are damaged almost equally by high-LET radiation, whereas it takes more low-LET radiation to kill hypoxic cells, perhaps because of differences in the number and type of damaging free radicals produced.

Free radicals, highly reactive and unstable molecules that contain an unpaired electron, have life spans of fractions of a second. Radiation can produce free radicals by interacting with oxygen, carbon, or water inside cells. Most X-rays damage cells as a result of the formation of hydroxyl radical ($\cdot$OH), which oxidizes nucleic acids in the cell and causes breaks in one or both strands of the DNA molecule. If both strands are broken, the cell dies because double-stranded damage is irreparable. On the other hand, mammalian cells have a high capacity for repairing single-stranded DNA damage, and this is a major natural defense against radiation injury.

Severe radiation damage results in immediate cell death. Less severe but still lethal radiation damage is expressed when the damaged cells attempt to divide and DNA must be copied (just before and during mitosis). When cells are radiated, breaks in the chromosomes occur, and broken ends can combine with the broken ends of different chromosomes during cell division. Such recombination, or crossover, events naturally increase genetic diversity by chromosome sorting in germ cells during meiosis, but in dividing somatic cells they may be lethal. This means that tissues that consist of numerous rapidly dividing cells, such as the epithelial lining of the intestines, are very sensitive to radiation damage. Total-body radiation exposures can be estimated by studying chromosomal damage in circulating white blood cells, the lymphocytes, and comparing it in vitro to cells exposed to a known dose of radiation. Lymphocyte analysis can detect a minimum body dose of about 0.1 to 0.2 sievert (10 to 20 rem).

A major biological effect of radiation is induction of cancer years after the exposure (stochastic effects). People who receive high radiation doses have a significantly higher risk of developing cancer twenty or more years after exposure because of persistent DNA mutations. The extent of functional recovery of a tissue is related to the number of progenitor, or stem, cells that remain alive after radiation. If stem cells are destroyed and not replaced from healthy tissues, radiation injury will persist. Late effects also develop independently despite recovery from acute radiation injury.

In all mammalian cells radiation exposure decreases survival rate. The mathematical relationship between radiation dose and the fraction of surviving cells has two components: linear and exponential. A typical survival curve for mammalian cells exposed to radiation is shown in Figure 19.1, in which the fraction of surviving cells is plotted on a semilogarithmic scale. For X or gamma rays, the dose–response curve has a shoulder that is followed by a straight line as dose is increased. The shoulder represents the cell's ability to repair sublethal injury. For alpha particles or low energy neutrons, the dose–response curve is a straight line from the origin. Thus, cell survival is an exponential function of dose.

Radiation is effective in cancer treatment when it causes greater harm to tumor cells than it does to normal cells. The biological responses to radiation exposure are summarized by the "four Rs." The first R, repair, includes enzymatic mechanisms for reversing intracellular injury. The second R, reoxygenation, refers to $O_2$ delivery to surviving cells after radiation injury has killed some cells. The third R, repopulation, is the ability of a cell population to divide and replace dead cells after an exposure. The last R refers to variability in radiosensitivity over a cell's reproductive cycle. Therapeutic radiation depends on exploiting differences in the four Rs between tumor and normal cells by delivering the radiation in appropriately timed increments or fractions.

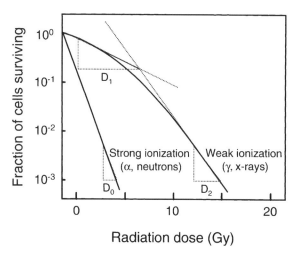

**Figure 19.1.** Survival of mammalian cells after exposure to radiation. The fraction of cells surviving (Y-axis) is plotted on a logarithmic scale against radiation dose (X-axis) on a linear scale. Strong ionization, alpha particles or low-energy neutrons, yields a straight line on a log-linear scale; cell survival is an exponential function of dose defined by the slope. Weak ionization, X-rays or gamma rays, yields a dose–response curve with an initial linear slope, a shoulder, and then at higher doses a straight line again with a greater slope. This curve is consistent with two components of radiation death: at low dose cell death is proportional to dose (D1); at high dose it is proportional to the square of the dose (D2).

Natural resistance to radiation is produced in some cells by a set of innate biological factors. These have been studied in detail in some tumors in which hypoxia and expression of tumor protection genes, or oncogenes, contribute to radiation resistance. However, a detailed conceptual understanding of radiation resistance remains problematic. This is an important area of research, both for the use of therapeutic radiation and to develop radioprotection strategies.

The response to radiation is also complicated by its induction of various growth factors and cytokines. For example, radiation induces expression of cytokines such as interleukin-1 (IL-1) and tumor necrosis factor (TNF). TNF and IL-1 protect blood cells from radiation, but TNF also enhances the killing of some human tumor cells by radiation. Thus, TNF may produce either radioprotection or sensitization depending on the type of cell involved. Part of a tissue's response to radiation includes elaboration of molecular growth factors from different types of cells. These growth factors, such as basic fibroblast growth factor (FGF), platelet-derived growth factor (PDGF), vascular endothelial growth factor (VEGF), and many others coordinate the response to injury but do not impart true adaptation or tolerance to additional radiation exposure. Some growth factors may actually exacerbate radiation damage to blood vessels and perpetuate inflammation. Others promote normal tissue repair but have also been implicated in scarring or fibrosis after radiation exposure.

## Radiation and the Human Body

Although a great deal is known about the cellular and molecular biology of radiation, most of what medical science knows about the human effects of total-body radiation is quite old. The information comes from the survivors of Hiroshima and Nagasaki in 1945, Pacific Islanders exposed to fall-out from U.S. atomic bomb tests between 1946 and 1958, and people exposed to radiation after the Chernobyl reactor accident in 1986. The world witnessed the devastating consequences of nuclear war on August 6, 1945, in the aftermath of the Hiroshima explosion in which a single bomb killed nearly 70,000 people and injured 100,000 more. Immediate survivors near the epicenter soon developed radiation sickness, and with the city's hospitals destroyed, effective medical care was impossible. This was followed a few days later by a similar catastrophe at Nagasaki, where thousands were left to die over the ensuing weeks and months.

Why do people die after exposure to radiation? The answer to the question is complex because different types of radiation are involved, and there is considerable variability in the total-body dose that humans can tolerate. People exposed to less than 2 grays generally require little, if any, therapy. If no medical support is available, the LD50/60 (the dose at which half the population is dead within 60 days) is about 3.25 grays. The very young and very old are more sensitive to radiation than are middle-aged adults, and women are more radiation tolerant than are men.

The clinical manifestations of radiation sickness depend on the total-body dose. The initial symptoms of acute total-body radiation, the prodromal radiation syndrome, last only a short time. At a dose of 100 grays or more, death from neurological and cardiovascular effects usually occurs within two days. This is known as the cerebrovascular syndrome. Because this syndrome causes death so quickly, other badly damaged body systems do not have time to fail.

At total-body doses of 10 grays to 20 grays death occurs from the gastrointestinal syndrome in three to ten days. Symptoms during this time may include days of nausea, vomiting, and diarrhea that lead to dehydration caused by sloughing of the intestinal lining (denudation). Septicemia and death soon follow. Most people exposed to more than 10 grays die from the gastrointestinal syndrome unless therapy (fluid and electrolytes, blood products, and antibiotics) is given. The extent of intestinal denudation depends on dose, and for exposures of 5 grays to 10 grays it may require a few days to several months. Intestinal denudation is usually fatal before the full effects on the bone marrow are seen.

At total-body doses of 2 grays to 8 grays death is caused by bone marrow failure, also known as the hematopoietic syndrome. It causes death three to six weeks after the exposure. The full effects of radiation on the bone marrow do not develop until the mature cells have been depleted, and symptoms may include chills, fatigue, and bleeding. Lymphocyte death and immune compromise occur in the first forty-eight hours after a significant radiation exposure. Death from infection or bleeding usually occurs before severe anemia develops because red blood cells have a long lifetime (roughly 100 days).

Victims who survive in the vicinity of a thermonuclear blast face many other health risks. Carcinogenic and genetic effects in atomic blast survivors are related to their distance from the epicenter of the explosion. After Hiroshima and Nagasaki, stable chromosome breaks and cancers were greatest in long-term survivors who were within about 2 kilometers (1 mile) of the epicenter (Schull, 1998). A detailed analysis of age-specific mortality rates for solid tumors as a group using the linear-quadratic dose-effect model of cell death revealed that the Hiroshima–Nagasaki data are consistent with a neutron RBE of about 70. This is within the accepted range, but it is about 3.5 times higher than the usual quality factor for neutrons. Eight types of radiation-induced cancer, leukemia, meningioma, thyroid, breast, lung, stomach, colon, and skin, have been described among these people, and more may be found as more survivors reach old age. In addition, there is evidence of increased cardiovascular mortality in elderly victims of radiation exposure (Fig. 19.2).

Medical studies also have provided estimates of the cancer toll of nuclear weapons testing. Today, these studies are relevant to the potential impact of radiation dispersal devices, or "dirty bombs," on an exposed population. These devices use conventional explosives to disperse radioactive material throughout a populated

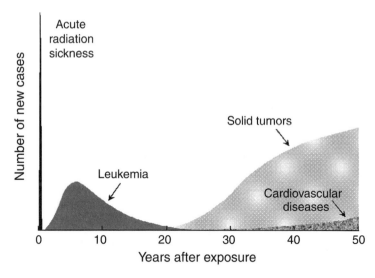

**Figure 19.2.** Appearance of radiation-induced diseases in a human population after exposure to a single thermonuclear detonation. Deaths from the effects of the blast and thermal injury are not included and would amount to most of the casualties.

area. Because the blast effect is relatively small, these weapons pose a more significant long-term than immediate risk to human health. Such devices, however, will only make effective weapons if they can disperse large quantities of highly radioactive material over population centers. Nonetheless, the health implications of dirty bombs can be appreciated using monitoring data for the effects of fallout from atomic bomb testing.

The Marshall Islands Nationwide Thyroid Study has regularly monitored the effects of the U.S. atomic weapons testing program conducted in the Pacific between 1946 and 1958. The nuclear tests contaminated a number of atolls with radioactivity and exposed Marshall Islanders to undefined levels of fallout. Between 1993 and 1997 the prevalence of both thyroid cancer and thyroid nodules was investigated among 4766 Marshall Islanders potentially exposed to radioactive iodine from fallout. The study found 68 thyroid cancers (and 13,498 benign nodules), a frequency of radiation-induced thyroid cancer of 1.4%. The National Cancer Institute has estimated that the release of iodine 131 in fallout from American nuclear test explosions has caused 49,000 excess cases of thyroid cancer in the United States. A 1991 study by International Physicians for the Prevention of Nuclear War estimated that the strontium 90, cesium 137, carbon 14, and plutonium 239 released worldwide in all nuclear test explosions would have led to 430,000 cancer deaths by 2000. In addition, there are concerns about the effects of nuclear weapons testing increasing the human germ cell mutation rate (Dubrova et al., 2002).

Further information is available from epidemiological analysis of Russian State Medical and Dosimetry Register data on more than 400,000 people exposed to radiation by the Chernobyl reactor accident, including 152,000 plant "liquidators" involved in cleaning up the site. The computed twenty-year risks of leukemia and solid cancers in liquidators are roughly 24% and 3%, respectively. A radiation epidemiological analysis of thyroid cancer in children made in Bryansk and Kaluga, downwind from Chernobyl, ascertained a relative risk of more than 7 for these cancers. The implication is that close to 90% of future thyroid cancers detected in the children of these regions will be attributable to Chernobyl radiation.

In summary, the major threat from weapons of mass destruction remains the detonation of thermonuclear devices over populated areas and the spread of radioactive fallout via the atmosphere. In the future, as the human population expands and nuclear weapons proliferate, their use can be expected to produce incomprehensible and ever-increasing devastation. To expect to survive a thermonuclear exchange between nuclear powers has long been known to be unrealistic because of the massive casualties and wholesale destruction of the medical and transportation infrastructures. The misguided philosophy of technoterrorism threatens to produce regional catastrophes of similar magnitude. The use of radiation dispersal devices also has serious implications for human health that are long lasting and not restricted to the site of detonation. For both types of devices, the scientific evidence clearly indicates that important carcinogenic and genetic effects of radiation will remain in those left alive near an epicenter as well as in many of their descendents.

# 20

## Human Prospects for Colonizing Space

Discussion of the prospects for human colonization of space is constrained by the scarcity of data on the biological and psychological effects of extended space travel. Limited observations are available on a small number of men and a tiny handful of women who have spent more than a month in space (see Fig. 18.1). This is a far cry from the extent of knowledge that will be needed if humanity is ever to steer a permanent course outside the friendly confines of Earth. The biological aspects of space travel have been largely neglected by futurists and science fiction writers, who assume that engineers will create artificial gravity, regenerative life-support systems, self-contained agricultural capabilities, and spacious environments that will enable people to live normally in deep space in an Earthlike environment. From an engineering standpoint, given an adequate supply of power, there is no reason, theoretically, to object to this vision of space colonization. However, the resources that such systems will require are enormous, and certain biological and psychosocial considerations go beyond engineering solutions. This is not to say that the engineering problems of life-support systems are trivial, for, as indicated in the following section, they are formidable, particularly when raw materials are unavailable.

## Advanced Life-Support Systems

A complete life-support system must provide, at a minimum, a breathable, uncontaminated atmosphere, a comfortable temperature, shielding from space and reactor radiation, adequate food and water, and resources to manage waste. Conceptually, life-support systems are of two types: open and closed systems. These systems are defined simply by whether the resources necessary to support life are recycled. Open systems supply everything but recycle nothing, while closed systems supply and recycle everything. Both types require external energy, but open systems require more resources and produce more waste. The resources required by an open system also increase in proportion to the length of time it is used. In a closed system an initial supply of resources is provided, then everything that is used is recovered and recycled.

In practical terms, current life-support systems used aboard submarines and on spacecraft are hybrid systems in which certain materials are recycled and others are consumed. Some nonrenewable resources are used only once; others are recycled in whole or in part for a certain period of time. In theory nothing prohibits a life-support system from being completely closed indefinitely, but the actual selection of open and closed technologies is determined by cost and practical considerations such size and weight. Purely closed systems are beyond today's technology because of flaws such as accumulation of useless by-products and leaks in the system. Also, the cost of recycling rises almost exponentially as closure approaches unity. Thus, even the capability of a nuclear submarine, which for all practical purposes has unlimited power and an unending source of water from which to generate oxygen, is limited by the supply of food available for the crew. A practical solution to the closed life-support system is a major hurdle that must be overcome for missions to Mars and beyond to be possible. Unlike the nuclear submarine, however, water will be limited, and the most appropriate source of power has not been decided.

Since the inception of the U.S. human space program with Project Mercury, life-support systems for low Earth orbit (LEO) and Moon missions have been primarily open because they are economically more practical for relatively short missions. The supply of oxygen, food, and water for the entire mission is on board at the launch, and essentially nothing is recycled. Aboard the International Space Station (ISS) the astronauts depend primarily on resupply except for the use of regenerative molecular sieves to absorb $CO_2$ and fuel cells to generate power from $H_2$ and $O_2$. Fuel cells can also produce potable water. Currently, some of the $CO_2$ generated by the crew on the ISS is also converted to water using the Sabatier process, a reaction discovered by the French chemist Paul Sabatier in the nineteenth century:

$$CO_2 + 4H_2 \rightarrow CH_4 + 2H_2O$$

The efficiency of this reaction is stimulated by a catalyst such as ruthenium. Although this technology for handling $CO_2$ is already quite advanced, an $H_2$ source is needed, and methane ($CH_4$) is currently discarded. Concentrated metabolic $CO_2$ can be dumped overboard, stored and recycled to make $O_2$ and water, or used later by photosynthetic plants in a bioregenerative ecosystem. There are also simple chemical processes that could salvage $CH_4$, which would allow recycling of valuable $H_2$ during future extended spaceflights.

The development of bioregenerative life-support systems depends on exploiting the ability of plants and other living organisms, such as bacteria, to sustain a closed ecosystem in space (Fig. 20.1). Indeed, the principle is the same as that for life on Earth itself, which has been a virtually closed, self-sustaining ecosystem for billions of years. However, the challenge for the future is to construct a highly reliable ecosystem on a comparatively tiny scale that is sufficiently redundant to tolerate the accidental loss of a large number of plants and bacteria from radiation, toxins, or diseases.

Substantial debate has centered on how much to invest in developing bioregenerative life-support systems or systems that are engineered entirely of artificial components. The outcome of this debate is not certain and will be resolved largely based on the solutions to the problems of producing, processing, preserv-

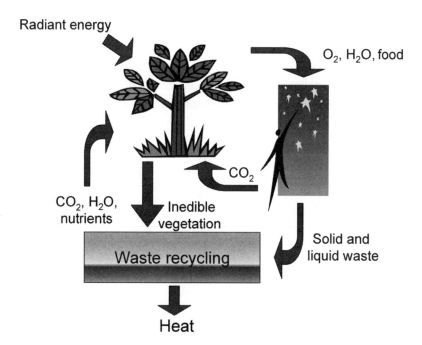

**Figure 20.1.** Elements of a bioregenerative ecosystem for extended spaceflight. These regenerative systems, once established, can theoretically operate much like the Earth simply with the input of energy and dissipation of excess heat to the outside.

ing, and storing enough food in space for extended spaceflights. Food production and $O_2$ generation are clear advantages of bioregenerative systems, and plants also have recognized value to the human sense of well-being. However, bioregenerative systems inherently suffer from problems such as slow response to changes in demand, amplification and drift of microbial populations, and problems with atmospheric and water contamination. Wholly artificial life-support systems hold theoretical advantages in terms of autonomy, reliability, and ease of maintenance. Currently, the science of synthesizing artificial foodstuffs from elemental components is in its early infancy.

## Mission to Mars

In 1996 NASA announced an ambitious series of unmanned missions to collect information needed for human beings eventually to travel safely to Mars. Robotic missions were planned, with goals ranging from detailed physical and chemical characterization of the surface of the planet to testing the equipment and technology needed to support the human journey and exploration of the planet. The purposes of the program are to define the hazards, design and test a vehicle prototype, and select a landing site. Future objectives are to determine the feasibility of living off specific resources that might be available in the Martian environment. In the language of NASA, this plan has been termed "In Situ Resource Utilization," or ISRU.

Natural resources are there for the taking on Mars, primarily water, which can be used directly or broken down to $H_2$ and $O_2$. The amount of water in the atmosphere is relatively miniscule, but small amounts of water ice have been detected in the polar caps. The recent mapping mission of the Mars *Odyssey* spacecraft indicated a surprising abundance of hydrogen in the top 3 feet of Martian soil, a sign of ground water ice even far from the poles, but water is nevertheless terribly rare relative to Earth soil. Martian soil composition has not been studied thoroughly, particularly in terms of the potential hazards of particulates and chemical reactivity. Its suitability also has not been ascertained for the construction of buildings or growing plants.

The thin Martian atmosphere is composed primarily of $CO_2$ (95%), $N_2$ (2.7%), and argon (1.6%). The $O_2$ concentration is 0.13%, which, at the atmospheric pressure of Mars, is less than 1 part in 17,500 of the $O_2$ in Earth's air. The concentrations of water and $O_2$ on Mars are far too low to be concentrated efficiently and inexpensively using transportable technology. ISRU thus remains largely hypothetical and will not be practical in the near future.

NASA's interest in ISRU for Mars missions implies that the problem of extended life support is, indeed, formidable. The NASA Mars Reference Mission of January 2014 to July 2016 would require thirty months for the round-trip. The scope of the problem is shown in Table 20.1, which indicates that a minimum of 25 tons of supplies, not counting containers and packaging or the fuel necessary to transport it, will be needed for a 900-day round-trip Mars mission for four people. The numbers in Table 20.1 can be used to argue against going directly to Mars without first establishing a near-Earth staging area. The weight of the supplies could possibly be reduced by a factor of two by recycling water, but more efficient and reliable technology must be developed and tested first. Prototype systems could be tested in LEO, such as on the ISS, or perhaps on the Moon, both of which, by virtue of proximity to Earth, provide logical test beds to set up and operate the advanced-life support systems that will be necessary for a Mars voyage. A round-trip mission to Mars stands a much better chance of success after a working prototype of the life-support system has been tested and its reliability and maintenance requirements determined.

Missions to Mars (or any other planet) will have to be designed to protect people and their equipment from an extremely hostile natural environment but must also protect that natural environment from people and their equipment. The natural environments of extraterrestrial worlds are no less vulnerable to damage than is the environment of this planet. In addition to adherence to the principles of conservation, there are important practical and scientific reasons to be circumspect about exploiting other worlds, including the risk of confounding the search for life outside our planet by contaminating the new environment.

On a journey to and from Mars, astronauts and their equipment will face the physical perils of ionizing cosmic radiation, solar particles, ultraviolet radiation, and impacts from meteorites and micrometeorites. At least initially, most of the trip will be undertaken in microgravity. On the planet the visitors will face the challenges of a rarified atmosphere, extremes of temperature, and great dust storms. The round-trip will require at least one year not counting the time spent exploring the planet, which will add another twelve to eighteen months to the trip. Most of the equipment needed for such an endeavor has yet to be invented, and without a major reorientation of American space policy, such an undertaking cannot realistically be expected until the middle of the twenty-first century.

The biological problems faced by astronauts on a trip to Mars will include, but are not limited to, loss of bone density, muscle atrophy, cardiovascular decondi-

Table 20.1. Supplies for a Round-Trip Voyage to Mars

| RESOURCE | REQUIREMENT PER ASTRONAUT (kg/DAY) | REQUIREMENT FOR 4 PERSONS FOR 900 DAYS (kg) |
|---|---|---|
| Oxygen | 0.75 | 2700 |
| Food | 0.75 | 2700 |
| Potable water | 2.5 | 9000 |
| Water for hygiene | 2.5 | 9000 |
| **Totals** | **6.5** | **23,400 (25 tons)** |

tioning, and ionizing cosmic radiation that may increase the risk of cancer. Inasmuch as the principles of relativity appear to be inescapable, Mars will be, by far, the simplest interplanetary journey that humans will ever undertake. Voyages to more remote locations within the solar system will require five to fifty times as long as will a trip to Mars and will be associated with more extensive compromise of the biological functions at risk.

## Habitability Factors in Long-Duration Spaceflight

To envisage what it will be like to live more or less permanently in deep space, engineers and behavioral scientists have studied environments on Earth that create similar physical and psychological stressors. These environments share the features of isolation, confinement, monotony, limited resources, and danger. Experiences relevant to space travel come from life aboard nuclear submarines, in underwater habitats, and at polar stations. However, submarine patrols last two to six months, and missions such as "wintering over" in Antarctica generally require isolation for a year or less. Most Earth-bound space simulation studies and ISS missions last only a few months, yet future interplanetary space travel will require years, if not decades.

In many confined environments, including ships, submarines, and prisons, human factors that have little to do with the immediate surroundings may have major effects on the perception of habitability. For instance, food is important not just for sustenance but because its quality and taste affect one's overall sense of well-being. Food can substitute, to some extent, for other quality-of-life factors, such as job satisfaction and social interactions. Almost nothing is known about the role meals might play as diversionary or social activities in long-duration space flight. Similar suggestions have been made for the roles of routine hygiene measures, such as hot showers and clean clothes, for which facilities even on the ISS are marginal. Regular exercise and various leisure activities will be essential in space for obvious reasons, but the effects of other life factors, such as air quality, humidity, noise, and odor have yet to be defined for extended periods in space.

Certain consequences of environmental stress can be attenuated if the exposure to the stressor occurs in a controlled manner. For example, it is less stressful to participate in a scheduled space walk than it is to have to put on a suit and go outside to investigate a possible micrometeorite impact on the water reservoir. This observation has two important implications. First, anticipated stresses should be scheduled for as many exposures as possible. Second, by understanding how planning counteracts stress, strategies can be developed to help individuals deal with it. Training enables one to match, insofar as possible, the individual with the environment in order to be reasonably sure that human capabilities will match conditions.

An important consideration in spacecraft design is that the ship will initially be home to at least four and later many more humans whose psychosocial needs will have to be met for a very long period, perhaps for the rest of their lives. How much and what kind of living space does this require? The requirements for quarters to live and work appear to vary with the duration of confinement. For a two-day confinement, it has been estimated that 50 cubic feet per person is the minimum, whereas for two months 150 to 260 cubic feet per person is needed; roughly seven hundred cubic feet per person is necessary for an indefinite stay. This seems like a lot of space compared to the capsules of the Mercury (55 cubic feet), Gemini (44 cubic feet), and Apollo (107 cubic feet) programs (see *Human Factors in Long Duration Spaceflight* 1972. Washington, D.C.: National Academy of Sciences), but the length and implications of the missions are not comparable. When completed, the 1-million-pound ISS will have a volume of 46,000 cubic feet!

The space per individual may vary with the number of individuals aboard, but it is unclear if more or less space per person will be needed as the size of the crew increases. This will likely depend on work schedules and the layout of work and recreational areas. Overcrowding obviously interferes with normal interactions and causes stress, which could alter physiological responses, create physical or psychiatric health problems, and degrade performance in space. The sense of crowding relates, to some extent, to one's frame of mind and personality. These factors affect perceptions of violations of privacy as well as expectations about what actions are appropriate to avoid them.

Individuals commonly respond to confinement by oversharing personal information with others but later may withdraw physically and psychologically from the group. This may reclaim some personal privacy but may also interfere with group dynamics. Privacy needs vary from individual to individual, and many of the usual means for maintaining privacy are absent in close environments. Anticipation of privacy needs will require more knowledge of why people maintain physical or psychological distance among themselves under confined conditions. Carefully selected and trained people are likely to adapt successfully to any reasonable environment, but it is not known how long this will take or how the adaptation will influence group dynamics.

Space use in microgravity theoretically should be more efficient because designers can work with the six surfaces of a cube. In the absence of the familiar cues of gravity, however, some astronauts prefer rooms with "ceilings and floors." It is also unclear how private and community space should be allocated and whether community areas should be flexible or dedicated to specific activities. People differ in their preferences for decor but are averse to environmental monotony and prefer diversity when work is light and as time passes.

One important problem of interplanetary space travel that appears to be insoluble will affect the outlook on the whole endeavor: communication between the spacecraft and Earth as the ship recedes from the solar system. After a few light years of travel, a huge information gap will be created between people as new knowledge accrued faster by those with more resources (presumably on Earth) will not be available to the others (presumably space travelers). After a few generations the events of the interstellar voyage and knowledge of human adaptation to space will be lost to Earth for the ensuing century. The "isolation" of space travelers from the rest of humankind and vice versa is not just a psychosocial problem; there are important biological implications for the survival of a small population that will be discussed at the end of the chapter.

## Deleterious Effects of Long-Term Exposure to Microgravity

If the adverse effects of microgravity turn out to be a major impediment to human colonization of space, then ships that generate gravitational fields will have to be used. On the other hand, spaceships that generate gravity depend on either acceleration or angular rotation, which restricts the living accommodations. Without portable gravity generators, livable space for colonists is reduced in a ship that spins or tumbles relative to a stationary ship of the same geometry. One of the best design ideas is also one of the oldest, the rotating wheel, which dates to Wernher von Braun in 1952. The advantages of the spinning wheel are that it provides a large, flat surface on the outer rim where the gravitational force is constant, and it minimizes the volume of atmosphere needed to support life by minimizing the distance between the inner and outer rims of the wheel. This set-up can be envisioned as a huge inflatable bicycle tire (or stack of tires) in which the voyagers live in the innertube and walk along the inside of the outer wall of the tire. The radius of the wheel will also have to be large enough to avoid motion sickness arising from the difference in centrifugal force between the head and feet of an upright person.

Generating artificial gravity in space assumes only the equivalence of planetary gravity and acceleration, in accordance with Einstein's relativity postulate. The biologist's perspective is that the need for artificial gravity in space is compelling because of the negative consequences of microgravity on bone and muscle mass, particularly because some of the changes may be irreversible. Of course, this assumes that the space travelers or their descendents intend to return to Earth some day or to colonize another planet with a comparable gravitational field.

What if the travelers intend to live permanently in space? Is there a rationale for allowing the voyagers to remain weightless and let adaptation or natural selection or both take their course? The answer to this question is unknown. The only extraterrestrial body likely to be colonized in the twenty-first century by humans is the Moon, which has one-sixth the gravity of Earth. Even for Mars, the only other site in the solar system likely to be visited this century, the argument to maintain gravity on a trip is not compelling given that the colonists eventually will have to adapt to a gravitational field only 38% that of Earth.

The decision to allow people to adapt permanently to microgravity cannot be made without the benefit of data on the consequences of prolonged weightlessness on the body. Such a decision would represent the beginning of a unique and lengthy experiment in the natural history of human biology. After hundreds of generations, numerous changes in the human phenotype would be likely as a result of weightlessness, radiation, and other differences in the environment. Such changes are extremely difficult to predict because of random factors in the genetics of small, closed populations. It is perhaps reasonable to suppose that natural selection would be diminished and that nonevolutionary pressures would have a greater impact on the voyagers. Indeed, this assumption has long been an operating paradigm of modern socioeconomic evolution. These issues will be discussed briefly at the end of the chapter, but one common speculation is that lack of natural selection in space would allow the body to revert to a more embryonic phenotype. For example, antigravity muscles might become vestigial, skeletal mass might decline, pigmentation might lessen, and the trunk might become smaller and the head larger. It seems plausible that manual dexterity and hand–eye coordination would be favored, and perhaps the legs and feet would become more useful for performing fine tasks.

## Effects of Life in Space on Human Immunity

Two other important and related consequences of living in space must be thought about in relation to multigenerational space travel, the effects of space on the immune system and the inexorable effects of exposure to various types of cosmic radiation on the human body. These topics are of interest on both practical and theoretical grounds, and there are scientific reasons to believe they interact through the effects of microgravity. In addition, radiation exposure directly interferes with the function of innate human immunity.

Any environmental stimulus that affects the nervous system will also affect the function of the immune system. The process is called neuroimmune modulation, or NIM. A wide range of physical environmental stressors has been associated with NIM, including heat, cold, electrical current, and changes in magnetic and gravitational fields (Spector et al., 1996). These physical parameters may affect either or both the cell-mediated, or humoral, arms of the immune system, thereby resulting in changes in susceptibility to infections, autoimmune diseases, and cancer.

In most instances NIM is mediated by cytokines and endocrine factors such as stress hormones, and the stress response is integrally involved. In-flight studies on astronauts suggest that stress-related hormones and other circulating factors, such as cytokines, suppress immune function in space. The cell type most clearly affected is the blood lymphocyte, which has become a model cell for space-based studies of human immune function. Impaired lymphocyte function in shuttle crews during and after spaceflight is shown by the depressed reactivity to stimulation of a subset of these cells of thymic origin called T-lymphocytes. T-lymphocytes respond to stimuli or mitogens, which normally provoke a robust response characterized by cell proliferation, but the responses of T-lymphocytes to stimulation by the mitogen *Con A* are reduced more than tenfold during and after exposure to microgravity.

If suppression of lymphocyte function in microgravity were due only to a stress response, then function could reasonably be expected to recover as the level of stress declined as well as perhaps show evidence of adaptation. However, part of the depressed immune response appears to be related to direct effects of microgravity on cell shape and function. Other factors, such as changes in circadian rhythm, may also contribute, particularly because the synchronization of external cues with the internal clock is disrupted in space. Unfortunately, the molecular mechanisms are not understood, which makes it impossible to know whether to expect adaptation.

To date, changes in lymphocyte function in microgravity have not been associated with specific clinical effects in astronauts, except, perhaps, for an increase in the tendency to acquire minor upper-respiratory tract infections. The situation is further complicated by lack of information on the effects of microgravity on the other arms of the immune system. These problems make it hard to venture even an educated guess about what will happen to the human immune system during extended space travel.

Despite the scarcity of scientific information, the possibility that extended space travel could change the human immune system permanently has some fascinating implications. One particularly interesting question is whether entirely self-contained space ecology offers an opportunity to leave behind the most troublesome, if not all, significant microbial pathogens. The image of the boy in the bubble comes to mind, wherein the child survives without innate immunity provided he remains sheltered from the bacteria and viruses of the outside world. Should permanent space travelers prevent any bacteria at all from accompanying them, even those that normally populate the skin and the gut? Or should scientists select or engineer some "allowable" strains? Realistically, the complete exclusion of bacteria and viruses for even a small number of individuals is neither practical nor reasonable, particularly if bioregenerative ecosystems are used, because they preserve the basic mechanisms for space-related changes in microbial growth, virulence, and host defenses.

Another fascinating aspect of the question has to do with what a lack of stimulation by pathogens would do to immune competence and immune surveillance. It is possible, although quite conjectural, that the coevolution of microbes with humans has been protective. Without bacteria and viruses, for instance, the occurrence of certain types of cancer might increase dramatically, particularly in the face of the expected long-term increase in radiation exposure and eventually result in the extinction of a small human population. This is essentially a restatement of Leigh van Valen's Red Queen hypothesis discussed in Chapter 10. On the other hand, bacteria and viruses cause many types of cancer and contribute to the pathogenesis of many others.

The surfaces of bacteria and viruses have highly discriminating systems to recognize and interact with molecules on the surfaces of living tissues. These recognition systems allow the microorganisms to attach to surfaces from which they would ordinarily be washed away by secretions such as tears or saliva and colonize the living epithelium. In bacteria the molecular lock-and-key mechanisms, or adhesin–receptor pairs, are highly specific both for the strain of organism and host as well as for specific types of surfaces in the host. In most cases the entire epithelium must be shed regularly to avoid overgrowth of microbes. For instance, intestinal turnover in germ-free animals is appreciably slower than it is in animals that live in a normal microbial environment. However, loss of too much epithelium is liable to result in infection of the deeper tissues. In addition, the host maintains many other defenses, such as antibody production, that modulate the adhesion process and regulate the inflammatory response.

The idea that infections prevent certain other diseases by stimulating the immune system is espoused in the *hygiene hypothesis*, which postulates that childhood infections impart life-long immunity to allergy (Yazdanbakhsh et al., 2002). It derives from studies in industrialized nations that link declines in childhood infection rates to increases in certain allergic illnesses. The hypothesis specifically attributes increases in asthma and diabetes to decreases in immune stimulation by fewer infections early in life. The proposal is that protective immunity is conferred by functional differences in T lymphocytes that develop after an infection. T lymphocytes occur naturally in two subsets known as T helper 1 (TH1) and T helper 2 (TH2) cells, which help regulate immunity by producing mediators that control allergic responses. According to the theory, childhood infections stimulate TH1 immunity, which opposes pro-allergic TH2 immunity. Less exposure to microbial pathogens weakens TH1 relative to TH2 immunity and increases the propensity to develop autoimmune diseases.

The hygiene hypothesis is not without problems: the evidence is inconclusive and marked by exceptions, such an increasing prevalence of TH1-mediated autoimmune diseases, and the effects of parasitic infections, such as pinworms, which stimulate TH2 immunity but are not associated with asthma. In addition, many other factors, such as genetics, predispose to asthma and diabetes. Thus, the hypothesis is undergoing revision. However, this does not negate the possibility that human immunity develops substantially under the influence of microbes. The point is that human interactions with microbes can be deleterious or advantageous and result in host–pathogen or host–symbiote relationships. Indeed, not only have host defenses against infection arisen, but many mutually beneficial relationships have developed on an evolutionary timescale. These relationships in a detached, artificially engineered ecosystem in deep space would be modified, lost, or degraded over many generations.

## Long-Term Effects of Radiation on Human Life in Space

Outside the Earth's atmosphere and magnetosphere, the flux of cosmic rays far exceeds that on the surface of the planet. The ionizing radiation above Earth's protective atmosphere is composed of a distribution of particles with energies as high as $10^6$ million electron volts (MeV) per particle. For humans in spacecraft and spacesuits, penetrating X-rays and gamma rays emitted from the Sun or generated by the interactions of charged particles with space hardware (*Bremsstrahlung*) are important sources of exposure. In addition, high-velocity subatomic and atomic particles derived from the Sun and other stars are important sources of radiation. The atomic flux is mostly (98%) protons and helium nuclei (alpha particles), but most of the particle energy is deposited by charged atomic nuclei larger than protons. These charged particles, or cosmic rays, are also known as high–atomic number energetic (HZE) particles. Because stars extrude elements only as large as iron ($Z = 26$), the flux of HZE particles larger than iron is small and primarily represents galactic radiation from novae and supernovae.

HZE particles can damage tissues by direct penetration, but, more importantly, smaller secondary particles, or spallation products, are generated when HZE particles impact a nucleus in the shielding or other hardware on a spacecraft. This radiation is the most important long-term health problem faced by astronauts, and the provision of appropriate shielding is its most obvious solution. In 1999 the Task Group on the Biological Effects of Space Radiation, formed by NASA and the U.S. National Research Council's Committee on Space Biology and Medicine, summarized current knowledge of the effects of long-term exposure to radiation in space and pointed out the need for more research in this area. Of particular importance was the imperative to determine the radiation-shielding constraints for interplanetary spacecraft (Setlow, 1999).

The biological responses of living organisms to ionizing radiation have been examined carefully for nearly a century, but much of this work has focused on the acute and chronic effects of controlled exposures, such as in radiation workers and on radiation as a therapy for cancer (see Chapter 19). Studies of cell proliferation, cell survival, gene expression, chromosomal damage, mutation rates, development, and longevity have been exhaustively conducted in ground-based laboratories. Space-based studies have been limited both in number and scope, particularly those intended to search for biological effects of exposure to natural cosmic radiation. Thus, the goal of understanding the long-term effects of naturally occurring cosmic radiation on human biology has so far gone unfulfilled.

The problem posed by exposure to cosmic radiation is illustrated by the following example. On a multiyear mission to Mars in a spacecraft with aluminum shielding of 4 grams per square centimeter, it has been estimated that up to 3% of the cell nuclei in an astronaut's body would sustain lethal damage from ionizing radiation, and the chromosomes of many other cells would be damaged permanently. This may have important immediate consequences to the central nervous system, including the brain, and would have definite consequences for developing cancer years after the mission. Simply adding more shielding to the Mars spacecraft is not practical because this adds weight, complexity, and cost to the vehicle. Furthermore, HZE particles that collide with nuclei in the shielding produce spallation products, which also have high energy and high linear energy transfer (LET) values. The best shielding materials, such as $H_2$, are lightweight and do not cause spallation. The problem of spacecraft shielding is therefore a classic of life-support system design that needs to be studied systematically to optimize the amount, type, and distribution of shielding for the vehicle. A plan to travel fast and then hide behind the planet, although it will reduce cosmic radiation exposure, is not sufficient. The thin Martian atmosphere offers little protection from cosmic radiation; thus, human visitors will need specific radioprotection measures.

The lack of data on the biological effects of HZE cosmic radiation is an important limitation not only for interplanetary spacecraft design but long-term human survival in space. The use of information on the biological effects of low LET radiation to predict the risks of HZE particles of high LET has not proven feasible because experimental connections are seriously lacking and extrapolating from low-LET to very high–LET conditions has inherent limitations. Cosmic radiation is not a problem of high dose rate but one of the cumulative effects of low radiation dose over months to years. The low dose rate produces a biological effect rate as much as tenfold lower than does acute exposure to the same dose. Therefore, uncertainty is produced largely by the biology, not by the physics, of radiation. Unfortunately, the time available to conduct the needed biological research at existing HZE accelerators is so limited that it could easily take another decade to determine the optimal shielding requirements for a lengthy Mars mission.

Protection against cosmic radiation, like any other radiation exposure, requires limiting the exposure to doses that minimize both direct (deterministic) effects and the risk of late (stochastic) effects after the dose. Deterministic effects follow standard dose–response relationships, while stochastic effects follow a probability distribution, which may be analyzed statistically but whose events cannot be predicted with certainty. The main stochastic risk of low–dose rate cosmic radiation is the increased incidence of cancer, for which there may be no threshold dose because theoretically a single HZE particle could produce a genetic modification in a specific cell of an individual that would lead to cancer. This process is called malignant transformation.

The probability of a dose of ionizing radiation killing a cell is about 10,000 times greater than the probability that the cell will become transformed, but to affect health the number of cells that must be killed is about 10,000 times the number that must be transformed to cause cancer. These two risks, cell death and transformation, therefore appear to be about equal, but the latent period for cell death is days to weeks, while the latent period for malignancy is ten to twenty years. Thus, one can argue about which is more important. The cancer mortality rate for U.S. astronauts selected from 1959 through 1991 is today not statistically different from the rates for age- and sex-matched case control subjects, but the at-risk population is still extremely small. Astronauts do, however, have a statistically higher mortality rate due to accidents and injuries, and those who have had higher radiation doses to the eyes may have an increased risk of cataracts.

One way to reduce the uncertainty about disease risk for an individual exposed to radiation is by means of biological dosimetry. Biological dosimetry is designed to assess the risk of long-term effects of radiation exposure by direct biological measurements. This differs from radiological dosimetry, which measures the dose of radiation that reaches the body, such as by using a thermoluminescent detector (TLD) or other counting device. The skin dose is then converted to an equivalent body or organ dose using an average radiation quality factor based on model calculations. Biological dosimetry is most commonly performed by cytogenetic analysis of white blood cells, such as lymphocytes. For many years measurement of radiation exposure in humans has relied on cytogenetic analysis of chromosome mutations in order to estimate the effects of absorbed doses.

Biological dosimetry has revealed a number of very interesting facts about humans. First, a great deal of interindividual variability exists in the background level of chromosomal aberrations, and certain aberrations, such as translocations, show particularly high natural levels. Recent cytogenetic analyses indicate that exposure to radiation is also associated with considerable variability in the induction of individual cellular responses. The extent of this variability depends somewhat on the method of analysis, but it also strongly differs with the mode of exposure, the doses delivered, and the type of radiation. Current estimates of the background frequencies of heritable risk in humans vary by at least two orders of magnitude, from approximately $10^{-5}$ to $10^{-3}$ per gene locus. In addition, the heterogeneity of particle radiation in the cosmos raises issues about the actual number of cells that will be damaged in an individual by a specific dose because densely ionizing particles are potent inducers of late chromosomal instability. This may have a significant effect on predicting the long-term risk of cancer.

The cumulative HZE exposure required to double the risk of cancer has been estimated to be at least 4 sieverts, which is more than the current NASA career exposure limits for astronauts allow (Table 20.2). The NASA limits were calcu-

**Table 20.2.** Astronauts' Career Limits—Whole Body Radiation Dose*

| AGE AT ENTRY | MALE | FEMALE |
| --- | --- | --- |
| 25 | 1.5 | 1.0 |
| 35 | 2.5 | 1.75 |
| 45 | 3.25 | 2.5 |
| 55 | 4.0 | 3.0 |

*Values in Sv

lated on the basis of an estimated 5% lifetime chance of contracting radiation-induced cancer. Because radiation-induced cancer takes years to develop, the probability of contracting such a cancer decreases with advancing age (or shorter life expectancy). Thus, in theory, an astronaut starting out at an older age can be allowed a higher lifetime exposure limit. This assumes, of course, that the latency period is independent of the age of the individual at the time of the exposure.

The biological effects of exposure to radiation on immune cells, as well as other cell types that divide rapidly, are an important independent concern for extended spaceflight. The cellular effects of radiation are also modified when microgravity is present. Although the force of gravity is too weak to have a significant effect on the deposition of radiation energy in tissues, secondary reactions that involve diffusion and convection of reactive chemical intermediates can be influenced by a lack of gravity. More importantly, cells and tissues sensitive to the influences of microgravity, radiation, or both recover more slowly after the cells are injured by radiation in the presence of microgravity than in normal gravity.

The influence of gravity on radiobiological processes may be understood in terms of processes that repair macromolecular damage. Many of these processes require the movement of enzymes or other molecules along oriented transport systems inside the cell. For example, the mitochondria, the powerhouses of the cell, migrate swiftly through the cytoplasm to bolster the supply of energy where it is needed most. These movements are guided to a large extent by a three-dimensional network of cytoskeletal elements called microtubules. Many repair processes require this scaffolding along with energy in the form of ATP to restore homeostasis. Thus, events that interfere with microtubule function or the metabolic status of the affected cell also interfere with cell repair mechanisms. Functional cytoskeletal elements are also needed to move and align chromosomes and orient the enzymes that repair DNA damage, such as strand breaks, in the nucleus of the cell.

Several types of experiments have been used to study the effects of radiation in space. Some have involved attempts to capture actual interactions of cosmic radiation with biological molecules, but because these are such rare events, the intentional exposure of cells or tissues grown in microgravity to controlled doses of radiation has been more informative to life scientists. Of human space flights to date, only the astronauts of the Apollo lunar missions have left the protective magnetic shroud of the Earth to be exposed to the cosmic radiation of deep space. Preliminary estimates from the Mars *Odyssey* spacecraft in 2001 indicated that the radiation exposure rate on a mission to Mars will be more than twice that which astronauts receive on the ISS orbiting 300 miles above Earth.

The literature is full of fascinating radiobiological studies that used seeds, spores, bacteria, beetles, and brine shrimp flown into space. These studies have proven the feasibility of germinating seeds and growing plants, and perhaps of farming animals, in space for food sources. One of the more interesting subjects of study has been the stick insect, *Carausius morosus*, which undergoes incomplete metamorphosis in the egg in 80 to 100 days. These organisms have been used to estimate the influence of microgravity on radiation-related changes in embryogenesis, development, maturation, and mutation rates. Studies involving mammals have been very limited. Although a number of radiobiological studies have been undertaken, the number of experimental organisms actually exposed to cosmic radiation in microgravity is quite small, which results in a paucity of interpretable data. Even so, several important concepts have emerged about the biological effects of naturally occurring radiation in space.

First, microgravity has a small but measurable effect on the biological response to radiation. The direction of the effect, however, is difficult to predict without actual reference to a specific endpoint. In other words, microgravity accelerates certain effects of radiation and delays others. Second, developing or metabolically active organisms are more sensitive to radiation in space than are adults and resting organisms. Of course, this is also true in ground-based studies, and the exact difference that spaceflight makes has not been decided. Third, the genetic lesions encountered most often in microgravity involve chromosomal damage or breakage. This finding strongly suggests that the interactions of microgravity with cell repair processes are going to be paramount in the effects of cosmic radiation on the body over the long run and may have effects on aging. The small number of individuals on any interstellar voyage will restrict the genetic diversity of the population, but the degree and effect of this are very difficult to estimate in the setting of long-term exposure to cosmic radiation.

## Establishment of Permanent Human Populations in Space

What should be expected of the lives of humans who reside permanently in space? Unfortunately, the scientific information is still far too sparse to make credible guesses. However, it is reasonable to expect that humans can adapt, survive, and reproduce in space. Permanent changes in the physical forces that act on the body, including the absence of sunlight and Earth's magnetic and gravitational fields, and the differences created by an artificial environment in external cues for synchronization of diurnal and circannual rhythms will almost certainly lead to some startling effects on the structure and function of the human body.

Given the magnitude of the difficulties that will be faced on a trip to Mars, which in interstellar terms is Earth's backyard, how does one plan for an expedition to the planetary system of a nearby star, during which many generations of people will have to live in deep space? The vehicle must obviously be large enough to carry an adequate supply of the essentials: power, oxygen, water, and nutrients. It must be capable of maintaining a stable atmosphere and temperature, supplying food and $O_2$, removing $CO_2$, and recycling other metabolic byproducts almost indefinitely. The craft must also have substantial redundancies, emergency systems, reserves, and repair capabilities. These life-support issues become major technological hurdles when one superimposes the substantial number of individuals that will be necessary to assure long-term stability of the gene pool in a small, closed population. Currently, these technological obstacles are far beyond our present capabilities. Nonetheless, it is informative to think about the human implications of such a voyage.

To appreciate the implications of isolating of a small group of people in deep space for many generations, some basic concepts of population genetics will be needed. A population is any group of individuals of the same species that share the same geographic location and who interbreed. The seminal tenet of population genetics is the Hardy-Weinberg principle (named after the two scientists who independently proposed it). The Hardy-Weinberg principle involves two critical bits of information: an estimate of the frequency of genes in a population (assuming random mating) and the prediction that the frequency of genes (alleles) in the population will remain at equilibrium given certain restrictions. Thus, Hardy-Weinberg predicts that the gene frequencies in a population will remain stable unless influenced by the forces of evolution. The Hardy-Weinberg equilibrium is a statement of a binomial theorem that demonstrates mathematically that the frequencies of genes transmitted in a predefined population are stable from generation to generation. This principle accounts for why dominant genes (such as brown eyes) do not propagate throughout a population and why recessive genes (such as blue eyes) do not die out.

To implement the binomial theorem, probabilities between 0 and 1 must be assigned to the frequencies of various alleles in the population. In organisms that reproduce sexually, each somatic cell is diploid. Its chromosomes contain genes with two alleles, that is, two copies of the same genetic information in the DNA, one from the mother and one from the father. Thus, humans have 23 pairs (46 total) of chromosomes. Individuals who carry the same information on both alleles of a gene are homozygous, and those who have different information on the two alleles are heterozygous. One of the two alleles is dominant over the other and is expressed preferentially in the phenotype. The silent allele is recessive, that is, it is not apparent in the phenotype of the carrier.

If the probability of a particular dominant allele A of a somatic gene (not on a sex chromosome) is p, then the probability of the recessive allele a is $1 - p$, denoted q. The sum of $p + q = 1$ because sperm and egg are haploid (one copy) and must contain either A or a, provided there have been no mutations. The reduction from a diploid to a haploid state in germs cells occurs before fertilization by the two-stage process of cell division called meiosis, which doubles the number of chromosomes but produces four daughter cells, each with exactly half the original number of chromosomes. The partitioning of genes is exactly equal unless there has been genetic recombination, in which a portion of a chromosome is exchanged during meiosis.

Because sperm and egg combine randomly, the possible genotypes and their frequencies are predictable. Given that the sum of the probabilities p and q is 1 in the haploid sperm or egg, the probabilities that the fertilized egg, or zygote, and hence the progeny, will have genotype AA, Aa, or aa are given by the binomial theorem: $p^2 + 2pq + q^2 = 1$ (see Table 20.3). In the progeny the probability of A is given by $p_p = f(AA) + \frac{1}{2} f(Aa)$, or $p^2 + \frac{1}{2}(2pq)$. Thus, $p_p = p^2 + pq$, and because $q = 1 - p$, then $p_p = p^2 + p(1 - p)$, or p. The frequency of A in the progeny is exactly equal to A in the parents! The same holds true for q and for the frequency of a. The binomial distribution proves that genes are shuffled, but the sum of the probabilities for p and q are each conserved. Thus, the overall gene frequency is stable from generation to generation.

To maintain the Hardy-Weinberg equilibrium, a population must satisfy five conditions: an infinitely large population (no genetic drift), no migration (no gene flow), no natural selection, no mutations, and no preselected mating (only random mating). If these five conditions are met, the genotypic (and allelic) frequencies in the population will remain the same from generation to generation. These restrictions and the equilibrium define the genetics of a nonevolving population. Deviations from the Hardy-Weinberg equilibrium are taken to indicate that one or more of the five conditions have been violated. Such deviations are necessary for evolution to occur.

The implications of the Hardy-Weinberg principle for human evolution in a permanent space colony are fascinating. The issues are most simply appreciated by considering the synthetic theory of evolution proposed by Sewell Wright, which attempts to explain evolution in terms of changes in gene frequencies. Wright theorized that a population evolves when gene frequencies change and thereby

**Table 20.3.** Gene Expression and the Binomial Theorem

| PHENOTYPE | GENOTYPE | FREQUENCY (f) |
|---|---|---|
| Dominant | AA | $p^2$ |
| Dominant | Aa | $2pq$ |
| Recessive | aa | $q^2$ |

improve the level of adaptation of the phenotype for a certain niche. Molecular forces of evolution, such as mutation and recombination, coupled with genetic isolation and migration of individuals with new alleles (gene flow) create natural heterogeneity in the population. Natural selection, the most powerful evolutionary force, then favors better-adapted individuals, and the population gradually evolves.

In a spacecraft far from Earth, both the population and the environment are closed, and one might expect the lack of gene flow and the lack of natural selection to delay evolution. On the other hand, mating could not be as random as it is on Earth, and mutation rates, both forward and backward, would be expected to increase because of exposure to cosmic radiation. These effects would tend to oppose the lack of gene flow and the absence of natural selection. The most pronounced genetic effects, however, would most likely derive from the small size of the population.

Population geneticists have long appreciated that multiple processes can create pronounced differences in the gene pool and the reproductive fitness of small populations. These include founder effects and the effects of population bottlenecks, inbreeding, and genetic drift. Founder and bottleneck effects describe similar events. Founder effects occur when a small number of individuals become isolated in a different location from the larger population. The derivative population reflects the main population only to the extent that its gene frequencies are a representative sample of the original population. Similarly, a small number of individuals can determine the gene frequencies in a large population if a drastic event dramatically reduces the size of a population, leaving behind a limited number of individuals to repopulate the environment. The survivors of these "bottlenecks" reproduce and expand the population. Although these survivors may have been chosen by natural selection, the selection process has temporarily reduced genetic diversity because most of the variability in the population was lost in the bottleneck.

In very small populations, random events can rapidly change allele frequencies independently of natural selection, mutation, or genetic recombination. This phenomenon is known as genetic drift. Because alleles are chosen at random instead of by natural selection or any other systematic mechanism, genetic drift has sometimes been called survival of the luckiest. As population size decreases, the drift becomes more pronounced and becomes stronger than natural selection. In addition, inbreeding between genetically related individuals tends to uncover deleterious recessive alleles that have been repressed by normal alleles, which allows expression of genetic disorders. For example, those who own highly inbred strains of dogs recognize that these animals are more likely to develop inherited disorders, such as hip dysplasia, and are generally less robust than are mixed breeds.

In small isolated populations deleterious alleles can become fixed, such as when only homozygous genotypes are expressed, within a few generations. The lack of heterozygous individuals reduces the overall genetic fitness of the population. Take the hypothetical case of a group of four families on a spacecraft in which all the parents are heterozygous (Aa) for a particular trait. The frequency of the a allele in these eight adults is eight per sixteen, or 50%. If each couple has four children, an average of four of the sixteen children would be expected to be homozygous dominant (AA), four homozygous recessive (aa), and eight heterozygous (Aa). The frequency of the a allele remains at 50%. If by random chance, however, the four families each have one child of AA and three children of Aa genotype, the frequency of the recessive allele in the group of children will fall from 50% to 37.5%. This genetic drift will have occurred in a single generation independently of natural selection or any other evolutionary pressure.

If the AA phenotype turned out to be deleterious to reproductive fitness in space, and if this random pattern continued for a few generations, the AA genotype might become fixed, thereby decreasing the genetic diversity and fitness of the population. This means a permanent human colony in space will have to be of sufficient size and genetic diversity to avoid random extinction. The minimum size necessary for a permanent, self-propagating space colony has never been determined, and this is indeed a difficult problem, particularly because there will be size constraints on the population due to limited space and resources. If the limited genetic diversity of reconstituted populations of endangered species is any indication, the minimum founder population for a remote permanent space colony is likely to be on the order of 100 to 200 unrelated individuals. Like an endangered species, close attention to the genetic composition of the population will be necessary.

In conclusion, the major biological concerns for colonization of space are the deleterious cellular effects of constant exposure to cosmic radiation and microgravity and the effects of physical confinement and genetic isolation on a small population of humans. On the basis of current information, these stressors might be expected to have negative consequences on skeletal density, physical strength, reproductive fitness, genetic fitness, innate immunity, and aging. The uncertainty in outcomes is less than an ideal way to embark on a dramatic new endeavor, but in many ways they are no less unpredictable than the future of life on Earth. On careful reflection, however, it would be unwise to underestimate the adaptability of the human organism, particularly with respect to the acquisition of knowledge and the unbridled capacity for behavioral adaptation when faced with new challenges.

# Bibliography and Supplemental Reading

Adolph, Edward F. 1947. Blood changes in dehydration. In: *Physiology of Man in the Desert*. Edited by Edward F. Adolph et al. New York: Interscience Publishers, pp. 160–171.

Adolph, Edward F. 1956. General and specific characteristics of physiological adaptations. *American Journal of Physiology* 184:18–28.

Ahlf, Peter, E. Cantwell, L. Ostrach, and A. Pline. 2000. Mars scientific investigations as a precursor for human exploration. *Acta Astronautica* 47: 535–545.

Alibek, Ken and Stephen Handelman. 1999. *Biohazard*. Random House: New York.

Anderson, G. S. 1999. Human morphology and temperature regulation. *International Journal of Biometeorology* 43: 99–109.

Aschoff, Jürgen and Rütger Wever. 1958. Kern und Schale im Warmehaushalt des Menschen. *Naturwissenchaften* 45: 477–485.

Auerbach, Paul S., ed. 2001. *Wilderness Medicine*, 4th ed. St Louis: Mosby.

Barcroft, Joseph. 1914. *The Respiratory Function of the Blood*. Cambridge, England. The University Press.

Blüher, Matthias, Barbara B. Kahn, and C. Ronald Kahn. 2003. Extended longevity in mice lacking the insulin receptor in adipose tiissue. *Science* 299: 572–573.

Behar, Michael. 2002. Defying gravity. *Scientific American* March: 286(3) 32–34.

Benzinger, Theodor H. 1969. Heat regulation: Homeostasis of central temperature in man. *Physiological Reviews* 49: 671–759.

Bennett, Peter B. 1993. Inert gas narcosis. In: *The Physiology and Medicine of Diving*, 4th ed. Edited by Peter B. Bennett and David H. Elliott. Philadelphia: W.B. Saunders, pp. 170–193.

Bennett, Peter B. and Jean Claude Rostain. 2003. High pressure nervous syndrome. In: *Bennett and Elliott's Physiology and Medicine of Diving*, 5th ed. Edited by Alf O. Brubakk and Tom S. Neuman. Edinburgh: Saunders (Elsevier Science), pp. 323–357.

Berson, David M., Felice A. Dunn, and Motoharu Takao. 2002. Phototransduction by retinal ganglion cells that set the circadian clock. *Science* 295: 1070–1082.

Bert, Paul. 1943. *Barometric Pressure: Researches in Experimental Physiology*. Translated by Mary Alice Hitchcock and Fred A. Hitchcock. Columbus, Ohio: College Book Co.

Bines J. 1999. Starvation and fasting: Biochemical aspects. In: *Encyclopedia of Human Nutrition*. Edited by Michèle J. Sadler, J. J. Strain, and Benjamin Caballero. New York: Academic Press, pp. 1779–1786.

Bouchama A. and J. P. Knochel. 2002. Heat stroke. *New England Journal of Medicine*, 346(25): 1978–1988.

Boycott, Arthur E., Guybon C. C. Damant, and John Scott Haldane. 1908. The prevention of compressed-air sickness. *Journal of Hygiene* 8: 342–443.

Brown, D. C. 1999. Submarine escape and rescue in today's Royal Navy. *Journal of the Royal Naval Medical Service* 85(3): 145–149.

Brubakk, Alf O., John W. Kanwisher, and Gunnar Sundnes, eds. 1986. *Diving in Animals and Men*. Trondheim, Norway: Tapir Publishing.

Butler, Patrick J. and David R. Jones. 1997. Physiology of diving of birds and mammals. *Physiological Reviews* 77(3): 837–894.

Cahill, George F. 1970. Starvation in man. *New England Journal of Medicine* 282: 668–675.

Cannon, Walter B. 1932. *The Wisdom of the Body*. New York: Norton.

Carson, Rachel. 1962. *Silent Spring*. Boston: Houghton-Mifflin.

Castro, Laura and Bruce A. Freeman, 2001. Reactive oxygen species in human health and disease. *Nutrition* 17: 161–165.

Center for the Study and Practice of Survival. 1997. *A Practical Guide to Lifeboat Survival*. Annapolis, Md.: Naval Institute Press.

Centers for Disease Control and Prevention. April 21, 2000. Biological and chemical terrorism: Strategic plan for preparedness and response recommendations of the CDC Strategic Planning Workgroup. *Morbidity and Mortality Weekly Reports* 49 (4): 1–14.

Charles, John B. 1999. Human health and performance aspects of Mars Design Reference Mission of July 1997. In: Proceedings of the First Biennial Space Biomedical Investigators' Workshop, League City, Texas, pp. 80–93. National Aeronautics and Space Administration, Johnson Space Center, Houston, TX.

Collins, Steve. 1996. The limit of human adaptation to starvation. *Nature Medicine* 1: 810–814.

Convertino, Victor A. 1996. Exercise and adaptation to microgravity environments. In: *Environmental Physiology*. Edited by Melvin J. Fregly and Clark M. Blatteis. New York: Oxford University Press, pp. 815–844.

Craighead, Frank C., Jr., and John J. Craighead. 1984. *How to Survive on Land or Sea*. Annapolis, Md.: Naval Institute Press.

Danzl, D.F. and R. S. Pozos. 1994. Current concepts: Accidental hypothermia. *New England Journal of Medicine* 331(26): 1756–1760.

DeHart, Roy L., ed. 1996. *Fundamentals of Aerospace Medicine*, 2nd ed. Baltimore: Williams & Wilkins.

de Onis, Mercedes, Edward A. Frongillo, and Monika Blössner. 2000. Is malnutrition declining? An analysis of changes in levels of child malnutrition since 1980. *Bulletin of the World Health Organization* 78: 1222–1233.

de Onis, Mercedes. 2000. Measuring nutritional status in relation to mortality. *Bulletin of the World Health Organization* 78: 1271–1274.

Do, N., H. LeMar, and H. Reed. 1996. Thyroid hormone responses to environmental cold exposure and seasonal change: A proposed model. *Endocrinology and Metabolism* 3: 7–16.

Dill, David Bruce. 1985. *The Hot Life of Man and Beast.* Springfield, Ill.: Charles C. Thomas.

Donald, Kenneth. 1992. *Oxygen and the Diver.* Flagstaff, Ariz.: Best Publishing.

Drew, Kelly L., Margaret E. Rice, Thomas B. Kuhn, and Mark A. Smith. 2001. Neuroprotective adaptions in hibernation: Therapuetic implications for ischemia-reperfusion, traumatic brain injury and neurodegenerative diseases. *Free Radical Biology and Medicine* 31(5): 563–573.

Dubos, René Jules. 1965. *Man Adapting.* New Haven, Conn.: Yale University Press.

Dubrova, Yuri E., Rakhmet I. Bersimbaev, Leila B. Djansugurova, Maira K. Tankimanova, Zaure Zh. Mamyrbaeva, Riitta Mustonen, Carita Lindholm, Maj Hultén, and Sisko Salomaa. 2002. Nuclear weapons tests and human germline mutation rate. *Science* 295: 1037–1051.

Edgerton, V. Reggie and Roland R. Roy. 1996. Neuromuscular adaptations to actual and simulated spaceflight. In: *Environmental Physiology.* Edited by Melvin J. Fregly and Clark M. Blatteis. New York: Oxford University Press, pp. 721–763.

Edmonds, Carl, Christopher Lowry, John Pennefather, and Robyn Walker. 2002. *Diving and Subaquatic Medicine,* 4th ed. London: Oxford University Press.

Ernsting, John, Anthony N. Nicholson, and David J. Rainford, eds. 1999. *Aviation Medicine,* 3rd ed. London: Butterworth-Heinemann.

Feist, D. D. and R. G. White. 1989. Terrestrial mammals in cold. In: *Advances in Comparative and Environmental Physiology,* vol. 4. Edited by Lawrence C. H. Wang. Berlin: Springer-Verlag, pp. 327–360.

Fitts, Robert H., Danny R. Riley, and Jeffrey J. Widrick. 2000. Physiology of a microgravity environment. *Journal of Applied Physiology* 89: 823–839.

Folz, Rodney J., Claude A. Piantadosi, and James D. Crapo. 1997. Oxygen toxicity. In: *The Lung: Scientific Foundations.* 2nd ed. Edited by Ronald G. Crystal, John B. West, Ewald R. Weibel, and Peter J. Barnes. Philadelphia: Lippincott-Raven, pp. 2713–2722.

Friedman, Jeffrey M. and Jeffrey L. Halaas. 1998. Leptin and the regulation of body weight in mammals. *Nature* 395: 763–770.

Gilbert, Daniel L. 1996. Evolutionary aspects of atmospheric oxygen and organisms. In: *Environmental Physiology.* Edited by Melvin J. Fregly and Clark M. Blatteis. New York: Oxford University Press, pp. 1059–1094.

Golden, Michael H. N. 2002. The development of concepts of malnutrition. *Journal of Nutrition* 132(7): 2117s–2122s.

Goldson, Edward. 1996. The effect of war on children. *Child Abuse & Neglect* 20(9): 809–819.

Grande, Francisco and Maurice B. Visscher, eds. 1967. *Claude Bernard and Experimental Medicine.* Cambridge, Mass.: Schenkman Publishing.

Guyton, Arthur C. and John E. Hall. 2000. Functional organization of the human body and control of the "internal environment." In: *Textbook of Medical Physiology,* 10th ed. Edited by Arthur C. Guyton and John E. Hall. Philadelphia: W.B. Saunders, pp. 2–8.

Hall, F. G. 1937. Adaptations of mammals to high altitudes. *Journal of Mammology* 18: 469–472.

Halliwell, Barry and John M. C. Gutteridge. 1999. Oxygen is poisonous—An introduction to oxygen toxicity and reactive oxygen species. In: *Free Radicals in Biology and*

*Medicine*, 3rd ed. Edited by Barry Halliwell and John M. C. Gutteridge. Oxford: Clarendon Press, pp. 1–35.

Hanák, P. and Petr Ježek. 2001. Mitochondrial uncoupling proteins and phylogenesis—UCP4 as the ancestral uncoupling protein. *FEBS Letters* 495: 137–141.

Harman, D. 1956. Aging: a theory based on free radical and radiation chemistry. *Journal of Gerontology* 11:298–300.

Hayward, J. S. and J. D. Eckerson. 1984. Physiological responses and survival time prediction for humans in ice water. *Aviation Space Environmental Medicine* 55: 206–212.

Hayward, J. S. and P. A. Lisson. 1992. Evolution of brown fat: Its absence in marsupials and monotremes. *Canadian Journal of Zoology* 70: 171–179.

Heacox, Kim. 1999. *Shackleton: The Antarctic Challenge*. Washington, D.C.: National Geographic.

Helfand, Ira, Lachlan Forrow, and Jaya Tiwari. 2002. Nuclear terrorism. *British Medical Journal* 324: 356–358.

Hensel, Herbert, Kurt Brück, and P. Raths. 1973. Homeothermic organisms. In: *Temperature and Life*. Edited by Herbert Precht, J. Christophersen, Herbert Hensel, and Walter Larcher. New York: Springer-Verlag, pp. 503–532.

Hensel, Herbert, Kurt Brück, and P. Raths. 1973. Homeothermic organisms: Heat exchange with the environment. In: *Temperature and Life*. Edited by Herbert Precht, J. Christophersen, Herbert Hensel, and Walter Larcher. New York: Springer-Verlag, pp. 545–564.

Hochachka, Peter W. and George N. Somero. 1984. *Biochemical Adaptation*. Princeton, N.J.: Princeton University Press.

Hochachka, Peter W. 1986. Defense strategies against hypoxia and hypothermia. *Science* 231: 234–241.

Houston, Charles S. 1960. Acute pulmonary edema of high altitude. *New England Journal of Medicine* 263: 478–480.

Hoyt, Reed W. and Arnold Honig. 1996. Body fluid and energy metabolism at high altitude. In: *Environmental Physiology*. Edited by Melvin J. Fregly and Clark M. Blatteis. New York: Oxford University Press, pp. 1277–1289.

Huey, Raymond B. and Xavier Eguskitza. 2000. Supplemental oxygen and death rates on Everest and K2. *Journal of the American Medical Association* 284: 181.

Huntford, Roland. 1999. *The Last Place on Earth: Scott and Amundsen's Race to the South Pole*. New York: Random House.

Huxley, R. R., B. B. Lloyd, Michael Goldacre, and H. A. W. Neil. 2000. Nutritional research in World War 2: The Oxford Nutrition Survey and its research potential 50 years later. *British Journal of Nutrition* 84(2): 247–251.

International Physicians for the Prevention of Nuclear War and the Institute for Energy and Environmental Research. 1991. *Radioactive Heaven and Earth: The Health and Environmental Effects of Nuclear Weapons Testing On and Above the Earth*. New York: Apex Press.

Jéquier, Eric. 2002. Leptin signaling, adiposity, and energy balance. *Annals of the New York Academy of Sciences* 967: 379–388.

Keatinge, William R. 1969. *Survival in Cold Water*. Oxford: Blackwell Scientific Publications.

Keatinge, William R., Gavin C. Donaldson, K. Bucher, E. Cordioli, L. Dardanoni, G. Jendritzky, K. Katsouyanni, A. E. Kunst, et al. 1997. Cold exposure and winter mortality from ischaemic heart disease, cerebrovascular disease, respiratory disease and all causes in warm and cold regions of Europe. *Lancet* 349: 1341–1346.

Keilin, David. 1966. *The History of Cell Respiration and Cytochrome*. London: Cambridge University Press.
Ladell, W. S. S. Water and salt (sodium chloride) intakes. 1965. In: *Physiology of Human Survival*. Edited by Otto Gustaf Edholm and Alfred Louis Bacharach. London: Academic Press, pp. 235–299.
Leaning, Jennifer and Langley Keyes, eds. 1984. *The Counterfeit Ark: Crisis Relocation for Nuclear War*. Cambridge, Mass.: Ballinger.
Logan, Nick and Dale Atkins. 1996. The snowy torrents: Avalanche accidents in the United States 1980–86. *Colorado Geological Survey* Special Publication 39.
Lukaski, Henry C. and Scott M. Smith. 1996. Effects of altered vitamin and mineral nutritional status on temperature regulation and thermogenesis in the cold. In: *Environmental Physiology*. Edited by Melvin J. Fregly and Clark M. Blatteis. New York: Oxford University Press, pp. 1437–1455.
Mack, Gary W. and Ethan R. Nadel. 1996. Body fluid balance during heat stress in humans. In: *Environmental Physiology*. Edited by Melvin J. Fregly and Clark M. Blatteis. New York: Oxford University Press, pp. 187–214.
Military Medical Operations Office. 1999. *Medical Management of Radiological Casualties Handbook*, 1st ed. Bethesda, Md.: Armed Forces Radiobiology Research Institute.
Merry, Brian J. 2000. Calorie restriction and age-related oxidative stress. *Annals of the New York Academy of Sciences* 908: 180–198.
Mersey, Lord John Charles Bigham. 1912. *Report of a Formal Investigation into the Circumstances Attending the Foundering on the 15th April 1912 of the British Steamship "Titanic" of Liverpool After Striking Ice In or Near Latitude 41 deg 46' N Longitude 50 de 14' North Atlantic Ocean, Whereby Loss of Life Ensued*. London: His Majesty's Stationery Office.
Mignot, E. 2001. A commentary on the neurobiology of the hypocretin/orexin system. *Neuropsychopharmacology* 25 (suppl 5): s5–s13.
Modell, Jerome H. 1993. Drowning. *New England Journal of Medicine* 328: 253.
Monge, Carlos C. 1943. Chronic mountain sickness. *Physiological Reviews* 23: 166–184.
Montain, S. J., W. A. Latska, and M. N. Sawka. 1999. Fluid replacement recommendations for training in hot weather. *Military Medicine* 164(7): 502–508.
Moore, Robert Y. 1997. Circadian rhythms: Neurobiology and clinical applications. *Annual Review of Medicine* 48: 253–266.
Mortola, Jacopo P. and Peter B. Frappell. 2000. Ventilatory responses to changes in temperature in mammals and other vertebrates. *Annual Review of Physiology* 62(1): 847–874.
Nakai S., T. Itoh, and T. Morimoto. 1999. Deaths from heat stroke in Japan: 1968–1994. *International Journal of Biometeorology* 43(3): 124–127.
National Heart, Lung and Blood Institute. 1998. *Clinical Guidelines on the Identification, Evaluation, and Treatment of Overweight and Obesity in Adults: The Evidence Report*. Bethesda, Md.: National Institutes of Health.
National Transportation Safety Board. Brief of Accident No. DCA00MA005 and CHI96IA157 and Air Accidents Investigation Branch Bulletin No. 6/99 Ref: EW/C98/8/6.
Nielsen S., J. Frokiaer, D. Marples, T. H. Kwon, P. Agre, M. A. Knepper. 2002. Aquaporins in the kidney: from molecules to medicine. *Physiological Reviews* 82(1): 205–44.
O'Hara, Bruce F., Fiona L. Watson, Hilary K. Srere, Himanshu Kumar, Steven W. Wiler, Susan K. Welch, Louise Bitting, H. Craig Heller, and Thomas S. Kilduff. 1999. Gene expression in the brain across the hibernation cycle. *Journal of Neuroscience* 19(10): 3781–3790.

Osborn, Mary J. and the Committee on Space Biology and Medicine, Space Studies Board, Commission on Physical Sciences, Mathematics, and Applications, National Research Council. 1999. A strategy for research in space biology and medicine in the new century. In: Proceedings of the First Biennial Space Biomedical Investigators' Workshop, League City, Texas, pp. 71–79. National Aeronautics and Space Administration, Johnson Space Center, Houston, TX.

Phillips, John L. 1998. *The Bends: Compressed Air in the History of Science, Diving and Engineering*. New Haven, Conn.: Yale University Press.

Piccard, August. 1997. My beautiful air-tight cabin. In: *From the Field: A Collection of Writings from National Geographic*. Edited by Charles McCarry. Washington, D.C.: National Geographic Society, pp. 106–108.

Pugh, L. G. C. E. 1965. High Altitudes. *Physiology of Human Survival*. Otto Gustaf Edholm and Alfred Louis Bacharach. London: Academic Press.

Ramey, C. A., D. N. Ramey and J.S. Hayward, 1987. The dive response of children in relation to cold water near drowning. *Journal of Applied Physiology* 63(2): 665–668.

Ramsey, J. D. and Thomas E. Bernard. 2000. Heat stress. In: *Patty's Industrial Hygiene*, 5th ed. Edited by Robert L. Harris. New York: Wiley, pp. 925–984.

Ravenhill, Thomas Holmes. 1913. Some experience of mountain sickness in the Andes. *Journal of Tropical Medicine and Hygiene* 16: 313–320.

Riedesel, Marvin L. and G. Edgar Folk, Jr. 1996. Estivation. In: *Environmental Physiology*. Edited by Melvin J. Fregly and Clark M. Blatteis. New York: Oxford University Press, pp. 279–284.

Roach, Robert C., Peter D. Wagner, and Peter H. Hackett. 2001. Hypoxia: From genes to the bedside. *Advances in Experimental Medicine and Biology* 502: 107–132.

Roberts, Jane C. 1996. Thermogenic responses to prolonged cold exposure: Birds and mammals. In: *Environmental Physiology*. Edited by Melvin J. Fregly and Clark M. Blatteis. New York: Oxford University Press, pp. 399–418.

Roff, Sue R. 1998. Puff the Magic Dragon: How our understanding of fallout, residual and induced radiation evolved over fifty years of nuclear weapons testing. *Medicine, Conflict & Survival* 14: 106–119.

Rowland, Neil E. 1996. Interplay of behavioral and physiological mechanisms in adaption. In: *Environmental Physiology*. Edited by Melvin J. Fregly and Clark M. Blatteis. New York: Oxford University Press, pp. 35–39.

Sakurai, Takeshi. 2002. Roles of orexins in regulation of feeding and wakefulness. *NeuroReport* 13(8): 987–995.

Schaefer, Karl E., H. J. Alvis, et al. 1949. Studies of oxygen toxicity: Preliminary report on underwater swimming while breathing oxygen. *Naval Submarine Medical Research Laboratory Report* 8: 84–93.

Schaefer, Karl E., ed. 1958. *Man's Dependence on the Earthly Atmosphere*. New London, Conn.: MacMillan.

Schmidt-Nielsen, Knut. 1981. Countercurrent systems in animals. *Scientific American* 244(5): 118–128.

Schmidt-Nielsen, Knut, E.C. Crawford, and Harold T. Hammel. 1981. Respiratory water loss in camels. *Proceedings of the Royal Society of London, Series B: Biological Sciences* 211(1184): 291–303.

Schmidt-Nielson, Knut. 1997. *Animal Physiology: Adaptation and Environment*, 5th ed. Cambridge: Cambridge University Press.

Schull, William J. 1998. The somatic effects of exposure to atomic radiation: The Japanese experience, 1947–1997. *Proceedings of the National Academy of Sciences* 95(10): 5437–5441.

Scott, Robert Falcon. 1996. *Scott's Last Expedition: The Journals*. New York: Carroll and Graf.

Setlow, Richard B. 1999. The U.S. National Research Council's views of radiation hazards in space. *Mutation Research* 430: 169–175.

Selye, Hans. 1993. History of the stress concept. In: *Handbook of Stress: Theoretical and Clinical Aspects*. 2nd ed. Edited by Leo Goldberger and Schlomo Breznitz. New York: The Free Press, pp. 7–17.

Shanley, Daryl P. and Thomas B. L. Kirkwood. 2000. Calorie restriction and aging: A life-history analysis. *Evolution* 54: 740–750.

Siegel, Jerome, Robert Moore, Thomas Thannickal, and Robert Nienhuis. 2001. A brief history of hypocretin/orexin and narcolepsy. *Neuropsychopharmacology* 25 (suppl 5): s14–s20.

Smoyer, Karen E. 1998. A comparative analysis of heat waves and associated mortality in St. Louis, Missouri—1980 and 1995. *International Journal of Biometeorology* 42(1): 44–50.

Smoyer, Karen E., Daniel G.C. Rainham, and Jared N. Hewko. 2000. Heat-Stress-Related Mortality in Five Cities in Southern Ontario—1980–1996. *International Journal of Biometeorology* 44(4): 190–197.

Sohal, Rajindar S., Robin J. Mockett, and William C. Orr. 2002. Mechanisms of aging: an appraisal of the oxidative stress hypothesis. *Free Radical Biology and Medicine* 33(5): 575–586.

Spector, Novera Herbert, Svetlana Dolina, Germaine Cornelissen, Franz Halberg, Branislav M. Marković, and Branislav D. Janković. 1996. Neuroimmunomodulation: Neuro-immune interactions with the environment. In: *Environmental Physiology*. Edited by Melvin J. Fregly and Clark M. Blatteis. New York: Oxford University Press, pp. 1537–1550.

Sridhar, K. R., J. E. Finn, and M. H. Kliss. 2000. In-situ resource utilization technologies for Mars life support systems. *Advances in Space Research* 25(2): 249–255.

Stolp, Bryant W., Claes E. G. Lundgren, and Claude A. Piantadosi. 1997. Diving and immersion. In: *The Lung: Scientific Foundations*. 2nd ed. Edited by Ronald G. Crystal, John B. West, Ewald R. Weibel, and Peter J. Barnes. Philadelphia: Lippincott-Raven, pp 2699–2712.

Swanson, Larry W. 1999. The neuroanatomy revolution of the 1970's and the hypothalamus. *Brain Research Bulletin* 50 (special ed., 5–6): 397–410.

Sweeney, James B. 1970. *A Pictorial History of Oceanographic Submersibles*. New York: Crown.

Taheri, Shahrad and Stephen Bloom. 2001. Orexins/hypocretins: Waking up the scientific world. *Clinical Endocrinology* 54(4): 421–429

Thompson, Ian. 1999. EVA dosimetry in manned spacecraft. *Mutation Research* 430: 203–209.

Tikuisis, Peter. 1997. Predicting survival time at sea based on observed body cooling rates. *Aviation Space Environmental Medicine* 68: 441–448.

Tomkins, Andrew. 2001. Nutrition and maternal morbidity and mortality. *British Journal of Nutrition* 85 (2 suppl): s93–s99.

Turner, Russell T. 2000. Physiology of a microgravity environment, invited review: What do we know about the effects of spaceflight on bone? *Journal of Applied Physiology* 89: 840–847.

U.S. Navy. 1999. *Diving Manual*, 2 vols. Rev 4, publ # 0910–LP–708–8000. Washington, D.C.: U.S. Naval Sea Systems Command.

Van Valen, Leigh. 1973. A new evolutionary law. *Evolutionary Theory* 1: 1–30.

Van Valen, Leigh. 1974. Predation and species diversity. *Journal of Theoretical Biology* 44(1): 19–21.

Ward, Michael P., James S. Milledge, and John B. West, eds. 2000. *High Altitude Medicine and Physiology*, 3rd ed. London: Arnold.

Webster, Donovan. 1999. Journey to the heart of the Sahara. *National Geographic* 195: 2–33.

Weibel, Ewald, R., ed. 1984. *The Pathway for Oxygen*. Cambridge, Mass.: Harvard University Press.

West, John B. 1984. Human physiology at extreme altitudes on Mount Everest. *Science* 223: 784–788.

West, John B. 2001. Historical aspects of the early Soviet/Russian manned space program. *Journal of Applied Physiology* 91: 1501–1511.

West, John B. 1998. *High Life: A History of High Altitude Physiology and Medicine*. New York: Oxford University Press.

West, John B. 1996. *Respiratory Physiology: People and Ideas*. New York: Oxford University Press.

West, John B. 1999. Barometric pressures on Everest: New data and physiological significance. *Journal of Applied Physiology* 86: 1062–1066.

West, John B. and Peter D. Wagner. 1980. Predicted gas exchange on the summit of Mt. Everest. *Respiration Physiology* 4: 1–16.

Yazdanbakhsh M., P. G. Kremsner, R. van Ree. 2002. Allergy, parasites, and the hygiene hypothesis. *Science* 296(5567): 490–4.

# Index

Aborigines, 108
Acceleration, 193–194, 195, 196, 197, 200, 204, 234
Acclimation, 16, 18–20, 55, 80, 106–108, 109, 205
  cross-acclimation, 21–22, 24, 28, 105, 148, 195
  de-acclimation, 16, 20
Acclimatization, 16–20, 24, 41, 50, 70–71, 75, 76, 105, 110, 169–170, 172, 175, 176, 185
  de-acclimatization, 20
Accommodation, 16, 19, 166
Acetazolamide, 175
Acute mountain sickness (AMS), 105, 173, 175
Adaptagents, 18
Adaptation, 3, 7–8, 13–14, 16–17, 22, 28, 41, 55, 65, 77, 80, 84, 95, 104–105, 106, 107–111, 112, 148, 149, 168, 172, 188, 194, 195, 200, 201, 206, 234
Adenosine
  diphosphate (ADP), 113–114
  monophosphate (AMP), 113
  triphosphate (ATP), 113, 170, 241
Adenylate cyclase, 49, 113

Adhesin-receptor pairs, 237
Adolf, E.F, 13, 50
Adrenal gland, 22, 23–24, 49, 195
Adrenocorticotropic hormone (ACTH), 23, 24
Afterdrop, 94
Aging, 136, 137, 139, 238
AIDS, 35
Aldosterone, 19, 70
Alkalosis, 168, 175
Allele, 243, 245–246
Allometry, 206
Alpha particles, 220, 223, 238
Altitude, 9, 18, 22, 24, 67, 99, 103, 105–106, 143–144, 164–170, 172–180, 184, 189–191
Altitude illness, 173–176
Alveolar air equation, 176
American Medical Research Expedition to Everest (AMREE), 179
Amino acids, 35, 61, 137
Ammonia, 129
Amundsen, Roald, 100, 102, 103
Andes, 172, 173
Androgen, 207
Android pattern, 38

## INDEX

Anemia, 170, 224
Anesthesia, 147, 151
Angiogenesis, 171
Antarctica, 99–100, 104–107, 232
Anthrax, 214, 217
Anti-diuretic hormone (ADH), 48
Anti-G suit, 200
Anti-gravity straining maneuver (AGSM), 200
Anti-oxidant, 34, 135
Apnea, 37
Apneusis, 59
Appetite, 22, 38, 39, 42
Aquaporin-2, 50
Archaea, 130
Arctic, 102, 106, 108, 119
Arctic fox, 65
Arginine vasopressin (AVP), 24, 48, 61
Armstrong Harry G., 188
Armstrong line, 188
Arrhenius relationship, 115
Arrhythmia, 42, 64, 83, 94, 122
Ascorbic acid, 17, 135
Asphyxia, 91, 159
Asthma, 237
Astronaut, 118, 191–192, 195, 208, 231–233, 240–241
Ataxia, 174
Atlas Mountains, 82
ATP synthase, 114
Atrial natriuretic peptide (ANP), 49
Avalanche, 91–92
Aviation, 11, 188, 190, 196, 200

Bacteria, 76, 96, 111, 130–131, 214–217, 229, 236–237
Balloonists, 184–185
Barcroft, Joseph, 11–12
Bar-headed goose, 180
Barents Sea, 152, 162
Basal metabolic rate (BMR), 107
Bathyscaphe, 184
Bay of Whales, 102
Bed rest, 205
Behavioral adaptation, 3, 7–8, 82, 84–88, 95, 97, 110, 112–116, 246
Bends, 141
Bennett, Peter, 141
Berber, 79, 82
Beriberi, 34, 36
Bernard, Claude, 10–11, 13
Bert, Paul, 11, 185
Beta particles, 220

Bicarbonate ($HCO_3^-$), 169, 175
Bilirubin, 43
Binomial theorem, 243, 244
Biological and chemical warfare, 213–214, 215, 216–217
Biological Weapons Convention, 214
Bioregenerative ecosystem, 229, 236
Birds, 60, 61, 63, 67, 82, 85, 87, 100, 107–108, 180, 193
Blackbody radiation, 66
Blackout (G-induced), 198
Blood plasma, 44, 45, 55
Blood pressure, 15, 16, 35, 49, 198, 199, 200
Blood–brain barrier (BBB), 174
Body fat, 32, 35, 37, 38, 42, 45, 105, 118, 122, 146, 174–175, 195
Body mass index (BMI), 31–32, 36, 37
Bond, George, 144
Bone, 46, 206, 208
Bone formation, 207, 208
Bone marrow, 32, 224
Bone mineral density (BMD), 207–208
Bone resorption, 207, 208
Bottleneck, 215, 245
Botulism, 214
Boyle's law, 142, 157
Bragg peak, 220
Breakpoint, 141
Breath-hold diving, 140
Bremsstrahlung, 238
Brown adipose tissue (BAT), 112–114
Brucellosis, 214
Built-in breathing system (BIBS), 159

Cachexia, 35
Caisson disease, 141
Calcium, 206, 207, 208
Calorie restriction, 138–139
Calorie requirements, 103–104, 118
Camel, 45, 79–83
Cancer, 37, 138, 221, 222, 224–226, 235, 239–241
Cannon, Walter B., 13
Carausius morosus, 242
Carbohydrates, 62, 139, 160
Carbon dioxide ($CO_2$), 62, 141, 143, 146, 151, 159, 160–161, 168, 176, 180, 183, 185, 191, 229, 243
Carbon monoxide (CO), 12–13, 159, 160
Cardiac output, 46, 48, 166–167, 198
Cardiovascular system, 36, 75, 77, 196, 198
Carotenoids, 135–136

Carotid body, 168, 170
Catabolism, 35, 104
Catalase, 135
Catecholamines, 24, 35
Cells
  adipocytes, 38, 39
  eukaryotes, 27, 130, 131
  germ cells, 17, 244
  myocytes, 94, 209
  neuroendocrine cells, 23
  osteoblast, 207
  osteoclast, 207
  osteocyte, 207
  prokaryotes, 129
  stem cells, 221
Centers for Disease Control (CDC), 213, 214
Cerebral blood flow (CBF), 174, 180, 199
Cerebrovascular syndrome, 224
Chaperone proteins, 27
Charles' law, 142
Chemoreceptor, 168
Chernobyl, 223, 226
Chemical weapons. *See* Biological and chemical warfare
Chilblain, 89
Childhood mortality, 31
Children, 30, 33, 43, 65, 74, 88, 90, 138
Chloride, 46, 55
Chlorophyll, 131
Chloroplasts, 131
Chromatin, 25
Chronic mountain sickness, 175–176
Circadian rhythms, 8, 105, 116, 236
Circulation, 11, 49, 51, 59, 142–143, 198, 205
Clo, 107
Coagulation, 77, 96
Cold, 20, 24
  habituation, 19, 105, 108, 109
  water, 7, 20, 68, 90, 100, 119–125
  weather, 66, 89, 95, 104, 147
Colley, Russell, 189–190
Comparator, 14
Compton effect, 220
Conduction, 66, 192
Continuity principle, 193
Continuous positive pressure breathing (CPPB), 200
Convection, 66–70, 192
Cooling, 67–68, 74, 75, 93, 95, 110, 122, 124, 127
Copper, 36, 98
Corticosteroids, 24
Cortisol, 35
Countermeasures, exercise, 206, 208, 210–211

Cosmic radiation, 238, 239, 245
Critical volume hypothesis, 147
Critical temperature, 20, 121
Cyanobacteria, 131
Cyclic guanosine monophosphate (GMP), 134
Cytokines, 25, 28, 35, 76–77, 223, 235


Dalton's law, 142, 165, 176, 185
Deceleration, 196–197
Decompression, 12, 142–143, 145
Decompression sickness, 140, 141–142, 143, 145, 189, 190, 191
Dehydration, 4, 6, 8, 28, 43, 46, 47, 51, 52, 54, 58, 60, 70, 75, 78–79, 80, 83–84, 86, 93, 94, 97, 122, 175, 198, 205
Desert, 3, 4, 6, 41, 52–53, 55, 78–84, 99, 111
Detector, 14
Diabetes, 15, 37, 38, 139, 237
Diabetes insipidus, 48
Diarrhea, 31, 35, 36, 43, 58, 224
Diuresis, 93, 96, 122, 205
Diving, 4, 6, 67, 125–126, 140–146
  mammals, 2, 140–141, 150
  reflex (response), 141
DNA (deoxyribonucleic acid), 26, 33, 38, 136, 137, 221, 241
Dosimetry, 219, 240
Dubos, Rene, 13


Ebullism, 188–189, 191
Edema, cerebral, 173, 174–175
Edema, famine, 31–34, 35, 139
Edema, pulmonary, 173–174, 175
Electrolyte, 35, 43, 50, 70, 77
Electron, 220
Elephant Island, 102
Embolism (air or gas), 141, 143, 155
Emergency position indicating radio beacon (EPIRB), 57
Endotoxin, 76
Endosymbiont, 131
Endothelial derived relaxing factor (EDRF), 134
*Endurance*, 101–102
Environment, 1–4, 8–9, 10–11, 13, 14, 16–20, 28, 50, 55, 59, 65, 72, 87, 105, 122, 129, 130, 140, 152, 181, 195, 230–231, 232, 235, 237, 242, 245
Epidemics, 29–30, 245
Epinephrine, 24
Error signal, 14, 16
Erythrocytosis, 169

## 258   INDEX

Erythropoietin (EPO), 170–171, 172
Estivation, 64, 111–112
Estrogen, 39, 207
Evaporation, 44, 50, 66, 70, 86, 192
Evolution, 13, 17, 86, 106, 110–111, 130–131, 235, 243–245
Exercise, 20, 45, 67, 68, 70, 71, 75, 103, 107–108, 118, 121, 124, 127, 164, 168, 172, 174, 180, 205, 207, 210–211, 232
Exosphere, 181
Extracellular fluid (ECF), 45, 47, 55, 60, 70
Extravehicular mobility unit (EMU), 191

Famine, 20, 29–30, 36, 138
Famine edema. *See* Edema
Fasting, 33–34
Fat. *See* Body fat
Fatigue, 19, 49, 94, 122, 173, 195, 200, 224
Fever, 25, 64
Fibroblast growth factor (FGF), 223
Fick equation, 166
Fish, 60, 61, 100, 150
Fission, 218
Fusion, 218
Fitness, 7, 9, 71
Folic acid, 33
Founder effect, 245
Franks, Wilbert, 200
Free radical, 27, 133–134, 221
Fresh water, 44, 61, 159
Frostbite, 66, 89–90
Frostnip, 89

G tolerance, 194, 197–200
Gamma-amino butyric acid (GABA), 147
Gagarin, Yuri, 203
Gamma ray, 220, 222, 238
Gas density, 149–150
Gastrointestinal syndrome, 244
Gene flow, 245
Genetic adaptation, 17–18
Genetic diversity, 215, 246
Genetic drift, 244, 245–246
Genetic recombination, 221, 244, 245
Genotype, 18, 130, 243, 244
Gills, 11, 100
Glanders, 213, 216
Global positioning systems (GPS), 57
Glucagon, 35
Gluconeogenesis, 33
Glucose, 15, 27, 33, 34, 35, 40, 115, 139, 160, 170, 195, 209

Glutathione, 34, 135, 139
Glycogen, 33, 35, 45
Gravity, 8, 193–194, 198, 203, 234, 241
Gray (Gy), 219
Gray out, 198
Group dynamics, 9, 105, 233
Growth retardation, 30–31, 32
Guanylate cyclase, 134
Gynoid pattern, 38

Habitat, 2, 106, 126, 232
Habeler, Peter, 22, 165, 185
Habituation, 16, 19, 105, 109–110
Haj, 83, 84
Haldane, J.S., 11–13, 49–50, 73, 143
Halley, Edmund, 141
Halophiles, 130
Hardy, J.D., 84
Hardy-Weinberg principle, 243
Harman, Denham, 136
Harvey, William, 11
Heat
  balance, 65–70, 74, 126
  capacity, 69, 126, 162
  cramps, 75
  escape lessening posture (HELP), 124
  escape activities, 82, 86–88
  exchange, 66
  exhaustion, 50, 75, 84
  loss, 67–68, 74, 94–95, 97, 124, 162, 192
  production (generation), 39, 67, 70, 73, 93, 107–108
  shock, 25
  shock factors (HSF), 26
  shock proteins (Hsp), 25
  storage, 67
  stress, 25, 66, 65, 72–73, 76, 84, 97
  transfer coefficient, 69, 121, 126
Heatstroke, 19, 50, 64, 73–77, 84, 88
Hematopoietic syndrome, 224
Hemoglobin, 11, 100, 166, 168–169, 174, 175, 199
Hemolysis, 82
Henry's law, 142
Heterozygous, 243, 246
Hibernation, 64, 111, 112–117
HIF-1 (hypoxia-inducible factor-1), 171
High altitude, 11, 24, 44, 67, 105, 143, 164, 166, 168–170, 176, 179–180, 185
High altitude cerebral edema (HACE), 173, 174–175
High altitude pulmonary edema (HAPE), 173–174, 175

High-pressure nervous syndrome (HPNS), 148–149
Hillary, Sir Edmund and Tenzing Norgay, 22
Himalayas, 18, 180
Hiroshima, 57, 143, 223, 224
Homeostasis, 8, 13–15, 63, 86, 106, 116, 123, 193, 198, 209
Homeothermic animals, 63
Homozygous, 243, 245
Hunger, 8, 35, 36, 38, 39, 40, 85, 104
Hydrogen, 129, 130, 183, 229, 230, 239
Hydrogen peroxide ($H_2O_2$), 133, 137
Hydrostatic indifference level (HIL), 205
Hydroxyl radical ('OH), 133, 221
Hygiene hypothesis, 237
Hyperbola, 2, 122, 201
Hypergravity, 193, 194–195
Hyperlipidemia, 37, 38
Hyperpyrexia, 75
Hyperthermia, 64, 73, 74, 84
Hyperventilation, 122, 141, 168, 179, 185
Hypocretin, 40
Hyponatremia, 50
Hypothalamic-pituitary-adrenal (HPA) axis, 22
Hypothalamus, 22, 23, 38, 40, 98, 115
Hypothermia, 38, 54, 58, 64, 84, 90–91, 92–96, 112, 120, 122, 125, 127–128, 159, 160, 162
Hypoxemia, 167–168, 174, 176
Hypoxia, 11–12, 22, 24, 141, 143, 144, 164, 167–168, 170, 172, 174, 176, 179–180, 185–188, 189, 191, 198
Hypoxic pulmonary vasoconstriction (HPV), 174
HZE particles, 238–239, 240

Icefish, 100
Ideal gas law, 142
Immersion, 4, 58, 66, 90, 119–125, 205
Immersion foot, 89
Immobilization, 207
Immunity, 35, 96, 224, 235–237, 246
In Situ Resource Utilization (ISRU), 230
Indianapolis, USS, 57–58
Inert gas, 142, 146, 147, 148
Infection, 25, 35, 36, 43, 76, 96–97, 224, 235, 237
Infectious diseases, 213–217
Infertility, 38, 39, 172–173
Inflammation, 25, 136, 223
Influenza, 96
Insects, 84
Insensible losses, 44, 52

Insulation, 81, 95, 106, 107, 108, 109, 123
Insulin, 34, 39, 114, 139
Interleukin (IL), 25, 35, 76, 77, 223
International Space Station (ISS), 8, 189, 228, 231, 232, 233, 242
International Standard Atmosphere, 166, 181
Intracellular fluid (ICF), 45, 47
Inuit, 108, 109
Iodine, 98, 225
Ionization, 219–220
Ionosphere (thermosphere), 181–183

James Caird, 102
Johnson Sea Link, 125, 162
Jolt, 197

K2, 178
Kalahari, 108
Keilin, David, 12
Kidney, 14, 35, 43, 47, 59, 60, 62, 70, 82, 169, 170
*Kursk* (K-141), 152–155, 157–158, 161
Kwashiorkor (protein deficiency), 34, 139

Lapps, 108
Latent heat of vaporization, 44, 67
Lateral hypothalamic area (LHA), 40
Law of constant extinction, 110
Leptin, 38, 39
Lethal factor, 217
Life support, 8, 9, 18, 106, 228, 229–230
 closed, 228
 open, 228
Lifeboat (raft), 54, 56, 119, 124
Limit physiology, 1, 110
Linear energy transfer (LET), 221, 239
Lipid peroxidation, 135–136
Lipids, 35, 135, 142, 147
Lipofuscin, 137
Lipopolysaccharide (LPS), 76–77
Lipostat, 39
Lithium hydroxide, 161, 162
Long QT syndrome, 122
Lung, 11–12, 44, 59, 77, 132–133, 137, 141–142, 146, 150, 166, 173–174, 176, 180, 185, 199, 200
Lymphocyte, 221, 236, 237, 240

Mach, Ernst, 194
Malaria, 31, 35

Malnutrition, 29, 31–32, 34, 93, 97, 138
Malthus, T. Robert, 29–30
Mammal, 2, 17, 22, 39, 42, 63, 65, 80, 82, 84, 86, 87, 107, 108, 111, 112–115, 138, 150, 151, 180, 194, 242
Marasmus (calorie deficiency), 34
Marianas Trench (Challenger Deep), 184
Marine mammals, 59–60, 119
Mars, 99, 185, 189, 191, 208, 230–232, 234, 239, 243
McCann bell, 157
Meiosis, 221, 244
Melatonin, 116
Mesosphere, 181, 182
Messner, Reinhard, 22, 165, 185
Metabolic equivalent (met), 107, 167
Metabolic rate, 33, 73, 107, 108, 118, 136
Metabolism, 31, 35, 38, 43, 62, 63, 84, 94, 107, 108, 109, 111–112, 115, 160, 166, 191
Methane ($CH_4$), 129, 229
Methanogens, 130
Microgravity, 193, 204, 206, 207, 208, 209–210, 233, 234, 236, 242
Mid upper arm circumference (MUAC), 31
Milieu-interieur, 10–11, 13, 42, 188
Miners and mines, 49–50
Mineralocorticoid hormones, 49
Mitochondria, 12, 98, 113–114, 131, 136–137, 139, 166, 167, 170, 209, 241
Mitosis, 221
Momson, Swede, 157, 162
Moon, 18, 231, 234
Motion sickness, 204
Mountain sickness, 106, 172, 173, 275
Mt. Everest, 11, 22, 165, 178–179, 185
Muscle (skeletal), 33, 35, 45, 68, 71, 104, 117, 267, 170, 179, 194–195, 205, 209–211, 231, 235
Mustard gas, 214, 215
Mutation, 12, 39, 136, 221, 240, 245
  genetic, 17
  somatic, 17
Myoglobin, 141, 209
Myosin, 209

Nagasaki, 143, 223, 224
Narcolepsy, 40
Narcosis, nitrogen (inert gas), 145–148, 149, 158
National Aeronautics and Space Administration (NASA), 190–191, 230, 238, 240
National Cancer Institute, 225

National Institute for Occupational Safety and Health (NIOSH), 73
National Institute of Aging, 139
National Institutes of Health, 37
National Transportation and Safety Board (NTSB), 186–187
Natriuresis, 205
Natural selection, 17, 18, 83, 107, 148, 235, 244–245
Negative feedback, 14–15, 16
Nephrons, 48
Neural tube defects, 33
Neurohypophysis, 48
Neuroimmune modulation (NIM), 235
Neuropeptides, 23, 24, 28, 40
Neurotransmission, 147
Neutron, 220, 222, 224
Niacin, 36
Niche, 2, 130
Nitric oxide (NO), 134, 135, 136
Nitrogen ($N_2$), 126, 129, 132, 142, 144, 146, 147, 149, 150, 158, 230
Nodal condensation probability, 5
Nomads, 79, 82–83
Nonshivering thermogenesis (NST), 107, 112–114
Norepinephrine, 24, 49, 113
North Pole, 95
Nototheniids, 100
Nuclear, biological and chemical (NBC) weapons, 218
Nuclear submarine. *See* Submarine
Nuclear winter, 219
Nutrition, 17, 31, 32, 34, 38, 95, 98, 102, 103, 207


Obesity, 20, 36–39, 139
Oncogenes, 223
Opiates, 116
Orexins, 40
Orthostasis, 35, 198, 204
Osmolarity, 46, 47–48, 51, 55, 58–59, 61, 82, 86
Osmoreceptor, 48
Osmotic pressure. *See* Pressure, osmotic
Osteoporosis (osteopenia), 207
Otolith, 204
Oxidative stress, 25, 34, 135, 137–139
Oxygen ($O_2$), 11, 129, 143, 144, 159, 164, 166, 175, 183, 185, 187–188, 189, 191, 229–230, 243
  cascade, 166
  consumption, 12, 166–167, 179

delivery, 166–167
partial pressure, 132, 165
toxicity, 132–133, 158, 160
Ozonosphere, 182


Pablo syndrome, 80
Pair production, 220
Pancreas, 15, 134
Panting, 66, 67, 86, 87
Parabiosis, 38
Parabolic flight, 205
Parameter, 14
Parathyroid hormone (PTH), 207
Pascal's law, 190
Pellagra, 36
Penguin, 86
Personal flotation devices (PFD), 6, 123
Phenotype, 17, 130, 243
Phosphorylation, 26
Photoreceptors, 116
Photosynthesis, 131
Physical diversity, 7, 20
Physiological adaptation, 16–18, 65
Physiological regulation, 13–16
Physiology, 1, 10–13, 38, 40, 42, 55, 92
Piccard
  Auguste, 183–184
  Jean, 183
  Jaccques, 184
Pikes Peak, 88
Pituitary, 23, 48
Plague, 213, 216
Planck's constant, 220
Plasma volume, 45, 46, 47, 49, 52, 70, 72, 75, 80, 169, 205
Plastids, 131
Platelet-derived growth factor (PDGF), 223
Plutonium, 225
Pneumonia, 31, 32, 35
Poikilothermic animals, 63
Polar T3 syndrome, 105
Polyakov, Valeri, 203
Polycythemia, 169, 175
Population genetics, 243
Positron, 220
Potable water, 41, 60, 61
Potassium, 34, 35, 36, 46, 55, 94, 159, 195, 218
Prather, Victor, 184
Pressure
  atmospheric, 150, 156, 158, 164–165, 185, 188–189

barometric, 17, 23, 27, 38, 99, 143, 164–166, 176, 211, 228, 260–261, 277, 279–280, 289
hydrostatic, 140, 148
hyperbaric, 132
osmotic (colloid), 34
Pressure reversal of anesthesia, 148
Pressure suit, 189–191
Primates (non-human), 67, 114
  chimpanzee, 194
  monkey, 139
progesterone, 39
Proprioception, 210
Protein, 25–27, 33–35, 38, 61, 62, 70, 76, 113–114, 135, 137, 139, 147, 160, 171, 209–210
Protein-calorie malnutrition (PCM), 34, 35
Protein synthesis, 25–26, 115, 210
Proteosome, 27
Protons, 113, 114, 220
Pyrogen, 25, 76


Q fever, 214


Radiation, 8, 21, 66, 67, 133, 135, 159, 192, 218, 219–226, 231, 238–242
  dispersal device (dirty bomb), 240
  ionizing, 21, 136, 218, 219–220, 232
  sickness, 224
  therapeutic, 222
Radioactive fallout, 218, 225
Radioprotection, 223
Radiosensitivity, 222
Radon, 218
Rapture of the deep, 145
Rate of living hypothesis, 136, 138
Ravenhill, Thomas, 173
Reactive nitrogen species (RNS), 135
Reactive oxygen species (ROS), 132, 133, 136, 138
Red Queen hypothesis, 110, 237
Relative biologic effectiveness (RBE), 221, 224
Rem (roentgen equivalent in man), 219
Renin-angiotensin system, 49
Reproduction, 16, 39
Respiration, 11–12, 131, 136, 170, 209
Respiratory heat loss, 127
Respiratory quotient (RQ), 160–161, 176–177
Respiratory water loss, 44, 45
Reticular formation, 147

Retinohypothalamic tract (RHT), 46
Riboflavin, 36
Ribosome, 27
Ricin, 214
Rickettsiae, 214
Risk, 5–6, 32, 36, 37–38, 91, 103, 121, 178, 197, 224, 226, 240
RNA (ribonucleic acid), 26, 27, 113, 171
Roentgen (R), 219
Ross Ice Shelf, 103
Ross, Malcolm, 184
Royal Navy (U.K.), 62, 123, 155, 163
Russian Northern Fleet, 152

Sabatier, Paul, 228
Safety of life at sea (SOLAS), 56–57
Sahara, 7, 42, 78–79, 83
Salt, 11, 34, 42–43, 47, 50, 55, 59, 60–61, 70, 83
Sarin, 215
Satiety, 38, 39
Schaefer, Karl E., 161
Scott, Robert F., 100, 102–104
Scurvy, 17, 36, 102
Search and rescue transponders (SART), 57
Seasickness, 60, 61
Seawater, 2, 43–44, 54–56, 60, 61, 121, 125, 140, 156
Second law of thermodynamics, 65, 126
Selye, Hans, 22
Semicircular canals, 204
Senescence. *See* Aging
Serotonin, 98
Set point, 14–15
Settling point, 15
Shackleton, Earnest, 100, 101–102
Shallow water blackout, 141
Shell and core thermal model, 68–69
Shielding, 228, 239
Shivering, 14, 20, 35, 68, 94, 96, 97, 103, 105, 107–109, 116, 121, 124, 126–127, 162
Shock (circulatory), 76–77, 93
Shuttlebox, 85
Sievert (Sv), 219
Skeletal adaptation, 206
Skylab, 194, 208
Sleep, 9, 22, 40, 94, 103, 114–115, 149
Sleep-disordered breathing, 37
Smallpox, 216
Smoking, 36, 136
Sodium, 42, 46, 47, 50, 52, 55, 60, 62, 70, 82, 94, 205
Sodium chloride, 42, 55, 60

Solubility (gas), 142
Sonora desert, 53
South Pole, 4, 95, 100–104, 106
Space sickness, 203, 204
Space travel, 117, 201, 227, 232, 236
Spacecraft, 105, 201, 211, 228, 238, 239, 245
Spaceflight, 117–118, 203–204, 205, 206, 209, 211, 229–230, 236, 242
Spacesuit, 4, 190, 191, 192, 238
Spallation products, 238–239
*Squalus*, SS, 157
Starvation, 4, 8, 29–36, 40, 44, 104
Stapp, John P., 197
Stochastic effects, 221, 239
Strain, 18, 20, 73, 122, 198, 200
Stratosphere, 165, 181–182, 183, 188, 189
Stress hormones, 22, 24, 35, 97, 105, 209, 235
Stress proteins, 25, 76
Stressor, 18, 19, 20, 21–22, 28, 94, 105, 232
Strontium, 225
Stunting, childhood, 30–31
Submarine, 58, 105, 125–126, 150, 152–163, 218, 228, 232
Submarine escape, 155–156, 163
Sudden disappearance syndrome, 122
Sunstroke, 75
Superoxide, 133, 136
Superoxide dismutase (SOD), 133, 135, 137
Suprachiasmatic nucleus (SCN), 40, 115
Survival analysis, 5–8
Survival time, 2, 7, 51–53, 61, 73, 91, 95, 103, 118, 121–123, 126, 128, 158, 160
Suspended animation, 117–118
Sweat gland, 19, 44, 70
Sweating, 19, 44, 53, 67, 70, 72, 74, 75, 80, 86, 87, 95
Synapse, 147, 150

T2–mycotoxin, 214
Taureg, 82
Temperature
  body temperature, 14, 19, 25, 32, 50, 52, 68, 70–75, 80, 85–87, 93, 97–98, 103, 107, 114–115, 122, 124
  core temperature, 20, 68, 92, 121
  critical temperature, 20, 121
  skin temperature, 14, 68–69, 72, 108
  wet bulb globe temperature, 73
Temperature regulation, 14, 65–66, 86, 93, 94, 98, 112
Terra Australis Incognita, 100

Terra Nova Expedition, 103
Test Ban Treaty, 218
Testosterone, 116
Thermogenesis, 25, 39, 107
Thermogenin (uncoupling protein, UCP), 113
Thermolysis, 86
Thermoneutral point, 72
Thermoneutral zone, 183–184, 126–127
Thermonuclear device (weapons), 212, 213, 217
Thermophiles, 131
Thermoregulation, 14, 75, 93, 94, 94–95, 111
Thiamine (vitamin B1), 34, 97–98
Thirst, 4, 8, 43, 46–47, 50, 61, 84, 85, 86
Thyroid hormone (T4), 25, 94, 97, 98
Thyroid releasing hormone (TRH), 98
Thyroid stimulating hormone or thyrotropin (TSH), 24, 105
*Titanic, RMS*, 119–120
Tocopherols (vitamin E), 135
Tolerance, 2, 19, 25, 28, 43, 73–74, 82, 105, 117, 148, 149, 194–195, 199, 200, 206
Torpor, 111–112
Toxins (and chemical poisons), 217
Transcription, 25–26, 171
Translation, 25–26
Trench foot, 89
*Trieste*, 184
Troposphere, 181–182, 189
Tuberculosis, 35, 216
Tularemia, 214
Tumor necrosis factor (TNF), 35, 76, 77, 223
Turtle, 136, 180

U.S. Air Force, 186, 188, 197
U.S. Army, 11, 89, 143
U.S. Navy, 57, 144, 155, 156, 159, 162–163, 184
U.S. Postal Service, 213
Ubiquitin, 27
Uncoupling proteins (UCP). *See also* Thermogenin, 113–114
Urea, 15, 43, 61–62
Urine, 43, 45, 47, 49, 55, 59, 61–62, 82, 97

Vaccines, 214, 215
Van Allen belts, 183
Van Valen, Leigh, 110, 237
Vascular endothelial growth factor (VEGF), 171, 223
Vasoconstriction, 68, 94, 108, 124, 126

Vasodilation, 68, 72, 174
Venezuelan equine encephalitis, 214
Ventilation, 62, 127–128, 141, 160, 168, 175–176, 188
Ventral medial hypothalamus (VMH), 38
Venus, 150–151
Viral hemorrhagic fever, 216
Virulence, 215, 217, 236
Viruses, 214, 216, 236
Vitamin
　A, 36, 135
　B1 (thiamine), 34, 97–98
　C (ascorbic acid), 17, 135
　D, 207
　E (tocopherol), 135–136
VX, 217

Waist circumference, 38
Walsh, Donald, 184
Warburg, Otto, 12
Wasting, 31, 35, 117, 210
Water
　channel, 49
　discipline, 50
　intoxication, 43
　poisoning, 50
　requirements, 41–47
　vapor, 44, 176, 188
Weapons of mass destruction, 212, 214, 226
Weight-for-height index, 31
Weightlessness, 94, 203, 209–210, 234
Werner's syndrome, 138
Wernicke-Korsakoff syndrome, 98
West, John, 179
Wind chill, 66, 68, 95
Winter, 22, 85, 89–90, 91, 96–97, 105, 111, 114, 117
Winterschlaf, 114
World Health Organization (WHO), 31
Wright, Sewell, 244

X-ray, 218, 220–221, 222, 238

Yellow fever, 216
Yellow rain, 214

Zeitgeber, 115–116
Zenith, 185
Zinc, 35, 36, 98
Zone of death, 176